Invertebrate Immunology

Sivakamavalli Jeyachandran
Balu Alagar Venmathi Maran

Invertebrate Immunology

Sivakamavalli Jeyachandran
Lab in Biotechnology & Bio signal
transduction, Saveetha Dental College
and Hospital, Department of Orthodontics
Saveetha Institute of Technical
and Medical Sciences
Chennai, Tamil Nadu, India

Balu Alagar Venmathi Maran (iD)
Organization for Marine Science and
Technology, Graduate School of Integrated
Science and Technology
Nagasaki University
Nagasaki, Nagasaki-ken, Japan

ISBN 978-981-95-1548-6   ISBN 978-981-95-1549-3   (eBook)
https://doi.org/10.1007/978-981-95-1549-3

This book is dedicated to the remarkable invertebrate species whose immune systems have been the focus of this work. More than 95 % of animals found on our planet are invertebrates, which have perfected highly efficient innate immune systems to thrive in pathogen-filled environments ranging from mountaintops to the deep sea. Their survival and their successful evolution based only on innate immunity epitomize the efficiency and adaptableness of the mechanisms of nature's defense. That is what this book is about: dedicating it to these extraordinary creatures as instruments of a deeper understanding of the whole of the immune systems in the animal kingdom.

We also dedicate this work to those researchers and scientists whose years of uncovering the subtleties of invertebrate immunity have served as foundational leaps for my own work. These findings have been very critical to progressing both essential and material immunology. Pioneering work from the duo has shed invaluable light on the evolutionary origins of immune mechanisms and how these systems are reshaping life on

Earth. Without their dedication and passion, this book would not have been possible.

We want to show our deep gratitude to our respective mentors who have had a great role in our academic path and in the research in invertebrate immunology. They've been essential to our development, wisdom, and support in spades. I thank our respective colleagues for the enrichment that continues in our work. They have broadened the scope of this book to encompass their insights into an interdisciplinary framework and made its general contribution relevant in the ongoing exploration of immune systems.

Finally, this book is dedicated to those who will follow us and continue to investigate this very intriguing invertebrate immunity. This is an ongoing journey of discovery in this field, and our hope is that this book will ignite curiosity and incite the next generations to continue to push our knowledge of innate immunity even further. The immune strategies of invertebrates are remarkable and provide valuable insights for future research that will have direct implications for health sciences, biotechnology, and pharmaceutical development.

We also dedicate this book to our respective family and loved ones, for their unbreakable support, to constantly keep us going. They have encouraged us in science and made the work to pursue this possible.

Finally, this book is a tribute to the invertebrates, whose immune systems teach us so much about the evolutionary history of immunity and which also have relevance to the future destiny of medicine and health sciences.

Sivakamavalli Jeyachandran
Balu Alagar Venmathi Maran

Preface

Although invertebrates constitute the vast majority of animal species on Earth, their immune systems have historically been overlooked in the study of vertebrate immunology, which encompasses both the innate and adaptive aspects. Vertebrates utilize a network of robust, antigen-specific immune responses and immunological memory, whereas invertebrates do not. This, however, is more than they rely on their highly developed and efficient innate immune system to be protected against a large array of pathogens. Therefore, a study of invertebrate immunity yields a special and most valuable insight into the early origins and significance of immune systems. I am an expert in the field of immunology and so am excited to share this in-depth examination of innate immunity in invertebrates. This book reviews in detail innate immune responses in these organisms with a focus on their early defenses against microbial invasion. In contrast to vertebrates, invertebrates do not possess the adaptive immune system, but have evolved extremely efficient means of rapid and immunologically specific recognition and elimination of pathogens independently of preexisting sensitization. In this book, the various components of invertebrate immunity are taken in depth: physical barriers, humoral responses, and cellular immunity, which underlie an effective defense protocol.

Pattern recognition receptors (PRRs), including but not limited to Toll-like receptors (TLRs), are one of the most remarkable aspects of invertebrate immunity, recognizing conserved structures present on pathogens that are absent in host cells. These receptors are a component of the body's ability to identify pathogen-associated molecular patterns (PAMPs) as essential precursors of the immune response. The book deals in detail with the activation of diverse signaling pathways that produce antimicrobial peptides (AMPs), lectins, and complement-like proteins to effect the destruction of pathogens, in response to the perception of PAMPs. In turn, these receptors have been able to evolve in invertebrates and thrive in pathogen-rich environments such as this, and the results of this research have shed new light on how ancient immune systems function. Evolutionary origins of invertebrate immunity are also explored in this book, scanning the development of broad immune responses from the earliest multicellular organisms. These events represent a seminal step in evolutionary progression for PRRs in invertebrates, enabling organisms to mount rapid defenses against a full range of pathogens. Knowing how these pathways evolved is critical to the understanding of vertebrate immunity since many of their molecular components are conserved in vertebrates. In the 1990s, the discovery of the toll receptor pathway in *Drosophila melanogaster* marked a breakthrough, as it was demonstrated, perhaps not surprisingly, that vertebrate immune systems might have evolved from primitive immune mechanisms found in early invertebrate lineages.

In addition, immune priming in invertebrates (a related immune phenomenon observed in invertebrates, analogous to immunological memory in vertebrates) is discussed, whereby this phenomenon enables invertebrates to boost its immune response during their re-exposure to pathogens. In a much less developed form (less developed than the adaptive immune system of vertebrates), this ability demonstrates the ingenuity of invertebrate evolution, showing that the absence of adaptive immunity need not preclude powerful innate immune responses. A fascinating example of the kind of way in which evolutionary pressures have molded immune responses in animals living in rapidly changing environmental conditions is the studies on immune priming of insects, crustaceans, and mollusks. A study of invertebrate immunity has vast practical implications. Along with giving us a sense of the development in an evolutionary sense of how immune systems have evolved, this area of research opens up immense potential for practical applications in the pharmaceutical and biotechnology industries. Many of the AMPs produced by invertebrates have potent antibacterial, antifungal, and antiviral properties, and invertebrate models are already contributing to the development of novel antimicrobial agents. In addition, immunostimulants originating from invertebrate immune systems are being investigated as a way to increase immune responses in aquaculture and veterinary medicine in those species that are unable to mount a functional adaptive immune system. Moreover, the study of invertebrate immunity provides important insights into the prevention and treatment of emerging infectious diseases and the global problem of antimicrobial resistance. Invertebrates inhabit a continually varying, pathogen-laden environment; the immune systems of invertebrates provide a wealth of natural defense mechanisms for combating antimicrobial-resistant strains of pathogens and improving human and veterinary health.

This book also finally points to the future in which they advocate for continued utility of invertebrate models for vaccine development, antimicrobial research, and immune health sciences. Evolutionarily significant and unsurpassed in diversity, invertebrates serve as informative models for immune function, and knowledge gleaned may have relevance to human and wildlife disease. In addition to being a good manual for immunologists, this book is an indispensable resource for anyone involved in evolutionary biology, pharmacology, and biotechnology wishing to understand the complexities of innate immunity. Therefore, the immune defense of invertebrates is one of the most ancient and efficient immune systems on Earth. Studying how these organisms avoid the ravages of pathogens in the absence of an "acquired" form of immunity teaches us critical principles of immune function, immune evolution, and development of new therapeutics. This book is aimed at providing an all-encompassing introduction to the fascinating world of invertebrate immunity, both its foundational knowledge and its cutting-edge research.

Chennai, Tamil Nadu, India   Sivakamavalli Jeyachandran
Nagasaki, Nagasaki-ken, Japan   Balu Alagar Venmathi Maran

Acknowledgments

I hope you all will enjoy this book and I felt supported throughout the process that brought it to completion. I could never have completed this work alone, as so many people played a part in contributing their expertise, giving me encouragement and being my never-ending source of support. I would first like to thank my mentors and advisors for their most useful mentoring and guidance toward not only my research but also my academic path. Immunomolecular events have played an important role in the development of this book by reason of their profound knowledge of immunology and their ability to inspire and challenge me. I am deeply grateful for their useful feedback, constructive criticism, and endless encouragement which have helped this work at every stage.

I owe a huge amount of thanks to my colleagues and research collaborators whose contribution to this book has been innumerable. Without their shared expertise and challenging discussions, this book would not have been possible. This project has largely been an affair of the collaborative spirit in the field of immunology. We are all grateful to them for their willingness to participate in thoughtful conversation, to share their knowledge, and to offer critical feedback that helps improve the quality of this book.

Thanks also go to the scientific community at large, for doing the important research on invertebrate immunity that has provided the seeds of some of the ideas in this book. Their endless pursuit of understanding the intricacies of what they were calling innate immune systems in invertebrates has been a source of continuous inspiration and has been an influence of my work. Their cutting-edge research has helped us to expand our understanding of the recognition and defense mechanisms against invaders by the immune system.

I also offer thanks to the publishers, who have provided me professional support, sustained patience, and a passion for quality which have meant that this book is now available to a greater readership. They're very experienced moving through the editorial and production process.

I owe my deepest thanks to my family. Their love and support have been constant. I keep working with their love and patience during the most trying times. This book would not have been possible without their belief in me. I owe my whole journey to them for their encouragement. Venmathi Maran thank Prof. Kavitha, Manikandarajan and Dhurga Harini for a great moral support in completing this task.

I would also like to recognize and acknowledge the invertebrate species themselves: their immune systems, hugely diverse and fascinating in themselves, have made this research possible and provided the insight into evolution and immune defense that led one to consider these model systems in the first place. To all of them, this book is dedicated and none of this research would have been worth making without them.

I hope that this work will provide a useful resource for students, researchers, and anybody interested in the evolutionary and practical implications of invertebrate innate immunity. I hope that it will continue to motivate new exploration and discovery of this fascinating field.

Sivakamavalli Jeyachandran
Balu Alagar Venmathi Maran

Contents

About the Authors

Sivakamavalli Jeyachandran is currently serving as an Associate Professor in the Department of Biotechnology at National College (Autonomous), India. She holds a PhD in Animal Health from Alagappa University and has a strong academic background, including an MSc in Biotechnology and a PGDBI in Bioinformatics from Bharathidasan University. Dr. Sivakamavalli has diverse research experience, including postdoctoral roles at National Cheng Kung University, Taiwan, and Bharathidasan University. Her research spans various fields, with significant contributions in molecular biology, bioinformatics, and marine science. She has been involved in several funded projects, including those supported by TARE, Conference & POWER grant, SERB, and DST, focusing on marine microorganisms, quorum sensing, and anti-biofilm activity. Dr. Sivakamavalli has received multiple awards, including the INSA Visiting Scientist Award (2022 and 2024) and the ICAR Hackathon Award (2022). In addition to her research, Dr. Sivakamavalli is an active participant in organizing conferences and workshops, having led 24 events. She has authored 80 research articles and 25 book chapters and contributed to numerous international journals. Dr. Sivakamavalli's professional influence extends to her role as a reviewer for 19 academic journals and her involvement in various advisory boards and collaborations with industry and academic institutions. She is passionate about educating and mentoring students, guiding PhD research, and contributing to the academic development of her field through numerous collaborations and initiatives.

Balu Alagar Venmathi Maran is currently serving as an Associate Professor at Nagasaki University, Japan. Prior to this, he held the position of Associate Professor at the Universiti Malaysia Sabah (UMS), for 6 years. He has also worked as a Scientist at the Korean Institute of Ocean Science and Technology (KIOST), Busan, South Korea, and as a Research Professor at Chonnam National University and Kyungpook National University in South Korea. Dr. Maran brings over 30 years of academic and research experience in the fields of marine science and aquaculture. His core expertise lies in the taxonomy of marine fish parasites and the use of natural products for parasite control in aquaculture. He has also made significant contributions to research on fish immune responses to parasitic infections, tetrodo-toxin analysis in parasites, jellyfish biodiversity, harmful jellyfish toxins, and the extraction and application of jellyfish-derived collagen in cosmetic and biomedical fields. Currently, Dr. Maran is engaged in an interdisciplinary marine science and technology project that utilizes artificial intelligence for biological imaging and data analysis, aiming to enhance the accuracy and efficiency of marine species identification and ecological monitoring. He has authored over 100 peer-reviewed research articles, contributed 10 book chapters, and edited 3 academic books. As principal investigator, he has led numerous nationally and internationally funded research projects focusing on marine biodiversity and parasitology. His research excellence has been recognized through multiple awards and gold medals for innovation and scientific contribution. Dr. Maran is actively involved in the academic community as an Editor for *PeerJ* (Q1) and *Frontiers in Marine Science* and as a Guest Editor for the *International Journal of Microbiology* (Wiley) and *Diversity* (MDPI).

Chapter 1
Advancements in Invertebrate Immunity

1.1 Introduction

Innate immunity serves as the first and most immediate line of defense in all living organisms, providing essential protection against a wide spectrum of microbial infections while maintaining homeostasis amid continuous exposure to environmental pathogens. This branch of the immune system is evolutionarily ancient and fundamentally conserved across metazoans. It operates through rapid, nonspecific mechanisms that detect and neutralize invading microbes, thus preventing the establishment and spread of infection (Bisola et al., 2024; Pull & McMahon, 2020). Remarkably, more than 95% of all animal species on Earth are invertebrates, which entirely lack the classical acquired or adaptive immune system characteristic of vertebrates. Unlike vertebrates, invertebrates do not develop immunological memory or antigen-specific immune responses mediated by specialized lymphocytes. Instead, their defense relies exclusively on a robust and sophisticated innate immune system composed of physical and chemical barriers, humoral responses, and cellular immunity (Müller et al., 2013; Sharrock & Sun, 2020). While inherently less diverse in molecular components compared to vertebrate immunity, the innate immune repertoire of invertebrates is nevertheless highly efficient and tailored for rapid recognition and elimination of pathogens, contributing to their evolutionary success in diverse habitats.

The innate immune responses in invertebrates involve multiple layers of defense. Physical barriers such as the cuticle and mucosal layers serve as the first obstacle to microbial invasion. Beyond these, invertebrates deploy soluble effector molecules including antimicrobial peptides (AMPs), lectins, and complement-like proteins that neutralize pathogens extracellularly. Cellular immunity is mediated by hemocytes, which perform phagocytosis, encapsulation, and production of reactive oxygen species to clear infections (Kaufmann et al., 2010; Loker & Bayne, 2018). Recognition of pathogens occurs through pattern recognition receptors (PRRs),

S. Jeyachandran, B. A. Venmathi Maran, *Invertebrate Immunology*,
https://doi.org/10.1007/978-981-95-1549-3_1

such as Toll-like receptors (TLRs) and peptidoglycan recognition proteins (PGRPs), which detect conserved pathogen-associated molecular patterns (PAMPs) common to microbial invaders (Janssens & Beyaert, 2003; Li & Wu, 2021; Steiner, 2004). Evolutionarily, innate immunity in invertebrates is of profound significance (Table 1.1). These organisms have persisted through hundreds of millions of years, exposed continuously to pathogenic challenges. The absence of an adaptive immune system has driven the refinement of innate immune strategies capable of mounting immediate and effective responses to infections without prior sensitization (Černý & Stříž, 2019; Iwasaki & Medzhitov, 2015). Such evolutionary pressure has resulted in mechanisms that balance efficient pathogen clearance with minimal immunopathology, ensuring survival in pathogen-rich environments. Indeed, this form of immunity exemplifies how ancient immune systems operate successfully despite lacking the clonal selection and memory features of adaptive immunity.

In vertebrates, the immune system comprises both innate and adaptive branches. Adaptive immunity, mediated by T and B lymphocytes, generates highly specific

Table 1.1 Comparative overview of innate immunity across invertebrate phyla

Phylum	Major immune cells	PRRs present	Key AMPs	Immune functions	References
Arthropoda	Hemocytes (granulocytes, Agranulocytes)	TLR, PGRP, C-type lectins	Cecropins, Defensins	Phagocytosis, encapsulation	Coates et al. (2022)
Mollusca	Granulocytes, Hyalinocytes	PGRP, lectins	Mytilins, Defensins	ROS production, Melanization	Steiner (2004)
Annelida	Coelomocytes	TLR-like, lectins	Lumbricin	Phagocytosis, agglutination	Cerenius and Söderhäll (2021)
Echinodermata	Phagocytes	Lectins, scavenger receptors	AMPs	ROS generation, encapsulation	Roy et al. (2020)
Nematoda	Granulocytes, Agranulocytes	PRRs (unc-93, tir-1)	Caenopores	Phagocytosis, AMP production	Pukkila-Worley et al. (2012)
Platyhelminthes	Hemocytes	Lectins, TLR-like	Schistosomins	Encapsulation, immune evasion	Young et al. (2012)
Cnidaria	Agranulocytes, Amoebocytes	TLR, lectins	Hydramacin	ROS, pathogen trapping	Parisi et al. (2020)
Porifera	Archaeocytes	TLR-like, lectins	Defensins-like	Pathogen neutralization	Müller et al. (2013)
Bryozoa	Coelomocytes	Lectin-like	Unknown AMPs	Pathogen recognition	Patnaik et al. (2024)
Tardigrada	Hemocytes	Putative PRRs	Putative AMPs	Environmental resilience, pathogen resistance	Schulenburg et al. (2004)

responses and immunological memory, allowing enhanced protection upon subsequent pathogen exposures. However, the adaptive response requires time to develop following initial infection, during which innate immunity provides critical immediate protection (Tosi, 2005). In contrast, invertebrates rely solely on their innate defenses, which compensate by being rapid and broadly reactive. Notably, some invertebrates exhibit phenomena akin to immune priming, where prior exposure to pathogens results in enhanced innate responses upon re-exposure, suggesting an intermediate form of immune memory (Melillo et al., 2018; Sułek et al., 2021). Despite differences, there exist striking commonalities between invertebrate and vertebrate innate immunity. Many molecular components of innate immunity are conserved, including TLR signaling pathways and NF-κB transcription factors. This conservation supports the hypothesis that vertebrate adaptive immunity evolved from primordial innate immune mechanisms present in ancestral metazoans (Zhang & Ghosh, 2001). Consequently, invertebrate immunity provides a powerful model to study fundamental principles of immune recognition, signal transduction, and effector functions, offering insights into the evolutionary origins of immunity. The study of invertebrate innate immunity has important practical implications. Understanding how these organisms defend against pathogens informs development of novel antimicrobial compounds and immunostimulants. Additionally, invertebrate models contribute to vaccine research, particularly for aquatic species in aquaculture, where innate immune responses can be leveraged to enhance disease resistance in the absence of classical adaptive immunity (Montagnani et al., 2024; Nguyen, 2024; Roy et al., 2020). Furthermore, exploring innate immune mechanisms offers potential strategies to combat emerging infectious diseases and antimicrobial resistance. In summary, innate immunity constitutes a highly efficient and evolutionarily ancient defense system fundamental to the survival of invertebrates, which represent the vast majority of animal diversity. Its ability to provide immediate and broad protection without the adaptive immune machinery exemplifies evolutionary innovation and robustness. Comparative studies of invertebrate and vertebrate immunity not only illuminate the origins of immune complexity but also provide valuable knowledge to advance veterinary, medical, and environmental health sciences.

1.2   Evolutionary Origins of Invertebrate Innate Immunity

The evolutionary history of innate immunity traces back to some of the earliest multicellular organisms, highlighting the fundamental need for defense mechanisms against environmental threats and microbial pathogens (Table 1.2). As life on Earth diversified, transitioning from simple unicellular forms to increasingly complex multicellular entities, the evolution of sophisticated immune defense systems became imperative. The rising complexity in organismal structure and physiology necessitated mechanisms that could efficiently recognize, respond to, and neutralize pathogenic invasions to ensure survival and reproductive success (Baluška et al.,

Table 1.2 Evolutionary milestones in innate immune mechanisms

Time period	Organism group	Key evolutionary development	Immune feature	Reference
Pre-Cambrian	Early metazoans	Emergence of PRRs	Basic pathogen recognition	Rathinam et al. (2024)
Cambrian	Arthropods	Toll signaling pathway	AMP induction	Imler (2014)
Ordovician	Mollusks	Diversification of lectins	Pathogen binding	Fujita (2002)
Silurian	Annelids	Expansion of coelomocytes	Phagocytosis	Cerenius and Söderhäll (2021)
Devonian	Echinoderms	ROS-based immunity	Pathogen oxidation	Buchmann (2014)
Carboniferous	Flatworms	Adaptive-like responses	Immune evasion	Young et al. (2012)
Permian	Cnidarians	TLR and lectin development	Signal transduction	Parisi et al. (2020)
Triassic	Porifera	Recognition proteins	Immune priming	Müller et al. (2013)
Jurassic	Nematodes	Innate immune signaling	TLR analogues	Pukkila-Worley et al. (2012)
Cretaceous	Tardigrades	Desiccation resistance mechanisms	Indirect immune protection	Schulenburg et al. (2004)

2023; Jack & Du Pasquier, 2019). Invertebrates, which represent some of the oldest and most diverse animal lineages, play a pivotal role in understanding the evolutionary trajectory of immunity. These organisms are considered a critical evolutionary stepping stone in the development of immune systems, providing insights into how ancient immune pathways evolved and diversified. The immune defense mechanisms present in invertebrates encompass deeply conserved molecular pathways shared with vertebrates, particularly within their innate immune systems. Such conservation underscores the antiquity and fundamental nature of these defense mechanisms (Rathinam et al., 2024).

One of the most crucial evolutionary advancements in immune recognition was the emergence of pattern recognition receptors (PRRs) in invertebrates. PRRs are specialized molecules that recognize pathogen-associated molecular patterns (PAMPs), which are conserved molecular motifs commonly found on the surface of microbes but absent in host organisms. The recognition of PAMPs by PRRs constitutes the initial and essential step in activating an immune response, enabling organisms to detect a broad range of pathogens through conserved structural features rather than species-specific antigens (Wicherska-Pawłowska et al., 2021). This strategy of immune recognition likely dates back to the earliest multicellular life forms, representing a conserved evolutionary tactic to discriminate self from non-self and mount an appropriate defense against microbial invasion (Buchmann, 2018). A landmark discovery in this evolutionary narrative was the identification of the Toll receptor pathway in the fruit fly *Drosophila melanogaster* in the 1990s. Originally identified for its role in embryonic development, the Toll receptor was later shown to play a central role in immune defense against bacterial and fungal pathogens (Imler, 2014). The Toll receptor in *Drosophila* triggers a signaling cascade leading to the activation of nuclear factor-kappa B (NF-κB)-like transcription factors and the subsequent production of antimicrobial peptides (AMPs), which serve as potent effectors in controlling infections (Cammarata-Mouchtouris et al., 2022; Schneider, 2022).

Importantly, the *Drosophila* Toll receptor shares homology with Toll-like receptors (TLRs) identified in vertebrates, highlighting a remarkable evolutionary link between the innate immune systems of invertebrates and vertebrates. Vertebrate TLRs function similarly by recognizing PAMPs and initiating immune signaling pathways that activate both innate and adaptive immune responses (Rauta et al., 2014). This conservation exemplifies how immune systems have evolved by adapting and modifying ancient signaling pathways to meet the demands of increasingly complex organisms. The Toll/TLR pathway represents one of the most significant milestones in immune evolution, demonstrating a shared molecular foundation for pathogen recognition across the animal kingdom (Voogdt & van Putten, 2016). Beyond the Toll pathway, other PRR families such as the peptidoglycan recognition proteins (PGRPs), scavenger receptors, and C-type lectin receptors have also been identified in both invertebrates and vertebrates, further emphasizing the conserved nature of innate immune recognition mechanisms (Fujita, 2002; Pees et al., 2015). These receptors collectively enable organisms to detect a wide variety of pathogens and initiate diverse immune effector responses, including phagocytosis, melanization, and the production of reactive oxygen species and cytokines (Eleftherianos et al., 2021) (Fig. 1.1).

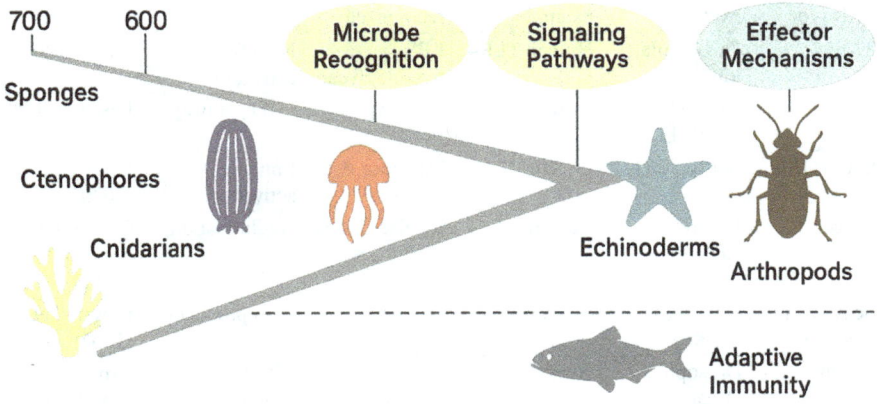

Fig. 1.1 Depiction of the phylogenetic timeline tracing the development of innate immune mechanisms in invertebrates and highlights evolutionary milestones marking the emergence of conserved immune components across major invertebrate taxa, illustrating how innate immunity predates adaptive immunity and is foundational to host defense

1.3   Humoral-Mediated Immunity in Invertebrates

While invertebrates lack an adaptive immune system, they have evolved a remarkably sophisticated humoral immune response that constitutes a fundamental aspect of their defense against invading pathogens. Unlike the antigen-specific memory and clonal expansion characteristic of vertebrate adaptive immunity, invertebrate humoral immunity relies on soluble immune factors present in body fluids such as the hemolymph in arthropods and other invertebrates that actively participate in pathogen neutralization and elimination (Boraschi et al., 2020). This humoral response is primarily mediated by the secretion of antimicrobial peptides (AMPs), proteins, and enzymes into the extracellular fluid, where they serve as potent agents against a broad spectrum of microbial invaders. The activation of this humoral defense system is tightly regulated and initiated upon pathogen detection by pattern recognition receptors (PRRs) expressed on immune cells or barrier tissues (Table 1.3). These receptors recognize conserved microbial molecules known as pathogen-associated molecular patterns (PAMPs), triggering downstream signaling pathways that coordinate the expression and release of humoral immune effectors. Importantly, the humoral response is often synergistically linked with cellular immunity where immune cells such as phagocytes work in concert with soluble factors to eliminate pathogens (McNeela & Mills, 2001). One of the hallmarks of invertebrate humoral immunity is the diversity and functional potency of AMPs. These peptides are typically small, cationic molecules exhibiting broad-spectrum

Table 1.3 Functional comparison: Pattern recognition receptors in invertebrates vs. vertebrates

PRR type	Invertebrate example	Vertebrate example	Ligand recognized	Downstream signaling	Reference
TLR	Drosophila toll	Human TLR4	LPS, peptidoglycan	NF-κB activation	Steiner (2004)
PGRP	Drosophila PGRP-LC	Absent	DAP-type PGN	IMD pathway	Iatsenko et al. (2016)
C-type lectin	Shrimp CTL	DC-SIGN	Mannose, Fucose	Immune activation	Wang et al. (2020)
Galectin	Earthworm galectin	Galectin-3	β-galactosides	Agglutination	Cerenius and Söderhäll (2021)
Scavenger receptor	Crustacean SR	SR-A	Oxidized LDL	ROS production	Frøystad et al. (2002)
Dscam	Drosophila Dscam1	Absent	Bacterial surfaces	Alternative splicing	Armitage et al. (2014)
Lectin receptor	Oyster lectins	CLEC4	Microbial glycans	Agglutination	Jia et al. (2016)
NLR-like protein	Cnidarian NLRs	NOD2	MDP	Inflammasome activation	Parisi et al. (2020)
FREP	Annelid FREPs	Fibrinogen	PAMPs	Opsonization	Romero et al. (2011)
βGRP	Shrimp βGRP	Absent	β-glucan	AMP induction	Amparyup et al. (2012)

antimicrobial activity against bacteria, fungi, and even certain viruses (Rodrigues et al., 2025). AMPs can exert their antimicrobial effects through multiple mechanisms: by disrupting microbial membranes, interfering with intracellular targets, or binding to pathogen surfaces to neutralize virulence factors. In the model organism *Drosophila melanogaster*, well-characterized AMPs include cecropins, defensins, and attacins. These peptides are rapidly upregulated upon infection and confer resistance to diverse pathogens (Hanson & Lemaitre, 2020). Their broad activity spectrum underscores the evolutionary advantage conferred by a rapid and flexible humoral immune arsenal.

The regulation of AMP production in invertebrates predominantly involves two major signaling pathways: the Toll pathway and the immune deficiency (Imd) pathway. The Toll pathway is primarily activated by fungal and Gram-positive bacterial infections, whereas the Imd pathway responds mainly to Gram-negative bacteria (Li & Xiang, 2013; Zhou et al., 2018). Both pathways converge on NF-κB-like transcription factors that induce the expression of genes encoding AMPs and other immune effectors, orchestrating a robust antimicrobial response (Caamaño & Hunter, 2002). Beyond AMPs, invertebrate humoral immunity incorporates a suite of additional soluble factors that contribute to pathogen neutralization. For example, phenoloxidase enzymes abundant in the hemolymph of arthropods catalyze the oxidation of phenolic substrates to reactive quinones, which polymerize into melanin around invading pathogens, effectively isolating and killing them (Coates et al., 2022). This melanization response not only provides a physical barrier but also generates cytotoxic intermediates that inhibit pathogen proliferation.

In addition, many invertebrates produce lectins, which are carbohydrate-binding proteins capable of recognizing specific sugar moieties on pathogen surfaces. Lectins participate in pathogen recognition, agglutination, and activation of other immune pathways, thereby facilitating rapid and efficient microbial clearance (Breitenbach Barroso Coelho et al., 2018). The collective action of AMPs, phenoloxidases, lectins, and other soluble factors forms a complex and dynamic humoral defense system that operates effectively in the absence of adaptive immunity. The evolutionary importance of such a rapid and broad-spectrum humoral response lies in its ability to provide immediate protection against a wide variety of pathogens without the delay required for adaptive immunity development. This immediacy is critical for invertebrates, which face continual exposure to microbial threats in their diverse and often microbially rich environments (Kulkarni et al., 2021). The extensive diversity of AMPs and other humoral factors observed across invertebrate species reflects the intense selective pressures these organisms have experienced throughout evolution. Adaptation to different ecological niches has driven the refinement of immune mechanisms tailored to the specific microbial challenges encountered, resulting in a highly efficient and versatile innate humoral immune system (Fig. 1.2).

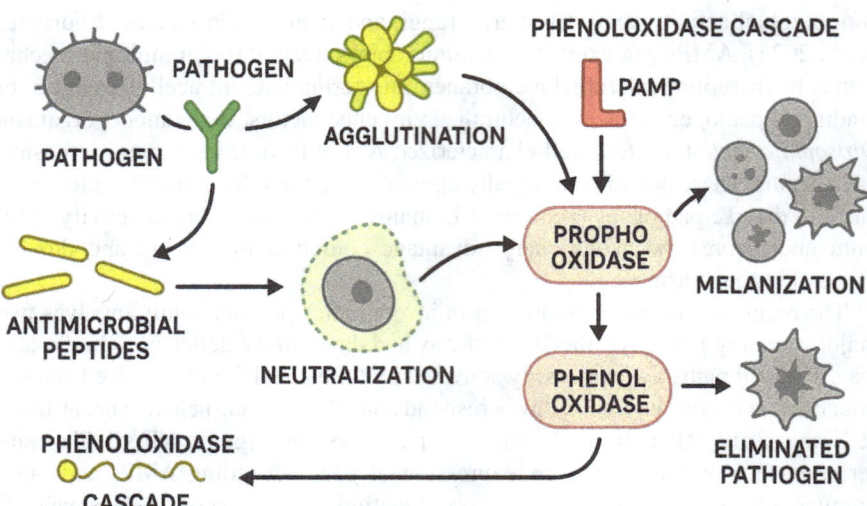

Fig. 1.2 The biochemical pathways and molecules involved in the humoral immune response of invertebrates. Soluble factors such as antimicrobial peptides, lectins, and the phenoloxidase cascade, showing their roles in pathogen recognition, neutralization, and elimination

1.4   Cellular Immunity in Invertebrates

Invertebrates exhibit cellular immunity alongside humoral responses to combat pathogens, showcasing a sophisticated and efficient defense mechanism. Hemocytes, the primary immune cells circulating in the hemolymph (the blood equivalent in vertebrates), play a central role in the invertebrate immune response. Hemocytes are responsible for pathogen recognition, engulfment through phagocytosis, and encapsulation of larger pathogens (Ciancio, 2016). The cellular immune system in invertebrates is highly effective, relying on pattern recognition receptors (PRRs) that detect pathogen-associated molecular patterns (PAMPs), which are conserved motifs found on the surface of pathogens (Patnaik et al., 2024). These receptors are crucial in initiating immune responses and engaging hemocytes to combat infections.

Phagocytosis, a key cellular immune response, is essential for eliminating pathogens. The surface receptors on hemocytes recognize and bind to PAMPs, triggering endocytosis, where the pathogen is engulfed by the hemocyte. Once engulfed, the hemocyte destroys the pathogen by releasing antimicrobial enzymes or reactive oxygen species (ROS), which serve to kill and degrade the pathogen (Coates et al., 2022). This process of pathogen elimination via phagocytosis is fundamental to invertebrate immunity and has been highly conserved across evolutionary time. The presence of these mechanisms in both invertebrates and vertebrates underscores the ancient origin of cellular immunity, with phagocytosis and ROS production forming the basis for more specialized immune responses in higher organisms. In addition to phagocytosis, encapsulation is another critical cellular immune response, particularly for pathogens too large for a single hemocyte to phagocytize. During encapsulation, multiple hemocytes aggregate around the pathogen to form a protective layer, effectively isolating the pathogen from the host tissue. This process is especially crucial for larger pathogens such as parasitic worms and bacteria. Encapsulation acts as an effective defense mechanism, ensuring that the pathogen is contained and unable to spread, thus preventing further infection (Liu et al., 2020; Loh et al., 2021).

The evolution of multicellularity in invertebrates is closely tied to the development of cellular immunity. As organisms became more complex, with specialized tissues and organs, the immune system required further advancements to defend against a wider range of pathogens. Hemocytes evolved to engage in various immune functions, including pathogen recognition, engulfment, and the signaling of immune responses. This evolutionary adaptation allowed invertebrates to develop a more advanced immune system capable of responding to diverse microbial threats in their environment (Rathinam et al., 2024). The conservation of cellular immune responses in invertebrates is significant because many of the same principles are found in vertebrate immune systems, albeit in a more specialized form. The basic principles of phagocytosis and encapsulation in invertebrates are echoed in the vertebrate immune system, where specialized cells like macrophages and neutrophils perform similar functions (Du Pasquier, 2001). This evolutionary conservation highlights the fundamental role of cellular immunity in immune defense across species and underscores the importance of studying invertebrate models to understand the origins and development of immune responses in animals. The importance of hemocytes in invertebrate immunity and their involvement in immune signaling further demonstrates the complex and integrated nature of cellular immunity. Recent studies have focused on the molecular pathways involved in hemocyte activation and signaling, revealing how hemocytes interact with other immune system components, such as antimicrobial peptides (AMPs) and complement-like systems (Huang & Ren, 2020; Zhao et al., 2023). The cross talk between cellular and humoral immunity in invertebrates provides valuable insights into how these systems work together to form an effective immune defense (Fig. 1.3).

1.5 The Evolution of Innate Immunity and Its Impact on Human Health

The immune systems of invertebrates offer many valuable insights into immune system evolution and the common principles that underlie immune defense, in general, across species (including humans). If the mechanisms of immunity are evolutionarily conserved in invertebrates and vertebrates, this would suggest that similarities hold with respect to principles of the immune system. In particular, this is clear in both invertebrates and vertebrates in the Toll signaling pathway and in the generation of antimicrobial peptides (AMPs), which constitute a critical arm of the innate immune response (Katzenback, 2015). In vertebrates, the Toll-like receptors

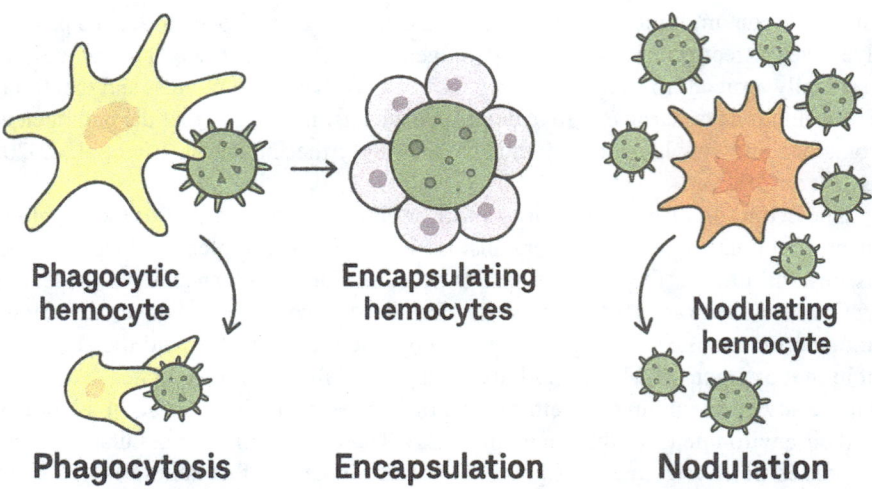

Fig. 1.3 Cellular components of invertebrate immunity, focusing on hemocytes and their functions like phagocytosis, encapsulation, and nodulation, and different hemocyte types, their morphology, and interactions with invading pathogens

(TLRs) were so named because they were first identified in invertebrates (e.g., Drosophila melanogaster) and act to recognize pathogen-associated molecular patterns (PAMPs), triggering the production of AMPs (Habib & Zhang, 2020). The discovery of these AMPs has made them important candidates for therapeutic development in the war against infectious diseases, as they exert a key role in pathogen elimination by killing or neutralizing bacteria, fungi, and viruses directly. Many invertebrates are dynamic models of immune function and provide important understanding of the evolution of immune systems. Invertebrate immune mechanisms are generally very highly conserved and thus provide an excellent model for the study how innate immune responses evolved in different taxa. A look at how immune systems of more complex organisms, through the evolutionary trajectory of invertebrate immunity, help shed light on how the immune systems of vertebrates might have evolved. Immune-related components such as TLRs, AMPs, and PRRs are functionally conserved, and vertebrate immune systems probably evolved from these ancestral immune tools (Nie et al., 2018). Studying invertebrate immunity, then, can inform fundamental principles of immune recognition, signaling, and pathogen defense—principles that are necessary to understand why and how the vertebrate immune systems, including human immunity, came to be (Cooper, 2010). Moreover, invertebrates are actively involved in human health not only as models for immune study, but also in their direct role in transmission of disease. In addition to their contribution to food security, several species of invertebrates are known vectors of human pathogens and are therefore critical to the public health landscape. Mosquitoes vector malaria and dengue fever; ticks are seen as vectors of Lyme disease and tick-borne encephalitis (Chikezie et al., 2024; Okoro et al., 2023). Invertebrates represent a valuable source of material to understand how organisms mount immune responses to infections, how pathogens interact with their hosts, and how immune responses are shaped in response to pathogen exposure. This knowledge can be leveraged for vector control strategies, e.g., to create immune-based intervention, whereby they reduce transmission of those pathogens or the ability of invertebrates to act as disease vectors (Hanson & Lemaitre, 2020).

Other than their role in disease transmission, invertebrates are a repository of novel antimicrobial agents. Antimicrobial peptides (AMPs) produced by invertebrates have been inspiration for new treatments to infections, including antimicrobial-resistant (AMR) infections. Because the global concern of antibiotic resistance is on the rise, the greatest need for new antimicrobials has never been greater. However, due to their broad spectrum activity and low potential for resistance development (Salam et al., 2023), AMPs, produced by a variety of invertebrates such as insects, crustaceans, and mollusks may provide a promising alternative. Since these peptides kill a broad range of microbial pathogens by disruption of their cell membranes, they are promising candidates in the development of new therapeutics against drug-resistant infections (Mba & Nweze, 2022). Pathogen recognition systems and the subsequent evolution of cellular and humoral immune responses in invertebrates are closely related to the evolution of invertebrate immunity. As organisms became more complex and invertebrates gave way to more complex organisms, their immune systems evolved to detect and neutralize a wider and widely more complex array of pathogens. In notable invertebrates including insects, crustaceans, and mollusks, highly sophisticated immune mechanisms were developed that permit the rapid detection and elimination of pathogens by phagocytosis, encapsulation, and production of AMPs (Rodrigues et al., 2025). These have essential roles in invertebrate immunity and their ability to survive in pathogen–rich environments, and invertebrate immunity continues to provide important insights into the evolution of the immune system and to suggest new therapeutic strategies. As genomic, proteomic, and metabolomic investigations of invertebrate immunity come into wider use, new pathways and mechanisms of immunity are being discovered. Increasingly powerful high-throughput sequencing technologies are rapidly disclosing new immune-related genes, signaling pathways, and immune response traits that could strongly influence the development and management of diseases, pest control, and the development of new antimicrobial agents (Nizamani et al., 2023). This continues to be the case the more we study the immune systems of invertebrates, which offer key insights into the origin of immunity and much promise for human, agricultural, and biotechnological health. With their evolved host defenses, honed by many millions of years of evolution, invertebrates represent an untapped resource of powerful tools for dealing with emerging infectious diseases and antimicrobial and vector-borne diseases that are relevant to basic and applied immunology.

1.6   Mechanisms of Innate and Adaptive Immune Response in Invertebrates

Because more than 95% of all animal species are invertebrates, and their immune systems have therefore been refined to act in this way, innate immunity is their most crucial form of immediate, broad-spectrum protection against pathogen activity. In contrast to vertebrates, invertebrates possess no adaptive memory of specific antigens or immunity (Silva & Gomes, 2024). However, as was recently reported, some invertebrates also have immune responses with qualities of adaptive immunity as well. For survival, these invertebrates utilize these mechanisms to fight pathogens in different environments. In invertebrates, there are several important key mechanisms of innate immunity, including pattern recognition, production of antimicrobial peptides, cellular responses, and tissue repair. The line of the first defense is made up of those physical barriers confining the pathogen entry (i.e., exoskeletons, cuticles, mucous membranes). However, if these barriers are damaged, pattern recognition receptors (PRR) of PAMPs activate signaling in the immune system. These receptors are for the recognition of conserved molecular structures of pathogens that lack these structures in the host and activate an immune reaction. One of the prime invertebrate PRRs that detect bacterial and fungal components and initiate responses of the immune system is the so-called Toll receptor, which was found in the fruit fly *D. melanogaster*. Furthermore, detection of pathogens also relies on other PRRs that are not Toll receptors, such as peptidoglycan recognition proteins (PGRPs) and C-type lectins (Wang et al., 2019a, b). These PRRs serve to recognize a pathogen, and following this recognition, a set of intracellular signaling pathways are activated that cause the production of antimicrobial peptides (AMPs). Small proteins are broad-spectrum antimicrobially active and neutralize bacteria, fungi, and viruses, a type, which are defensins, cecropins, and attacins, AMPs. As a defense mechanism to infections, invertebrates have developed a utilitarian immune response against pathogens that includes the production of AMPs by the host in response to infection, which are key elements to the humoral arm of the innate immune system. As far as banded hemocytes at the cellular level go, these are equivalent to macrophages in vertebrates. Hemocytes perform the process of phagocytosis, wherein the hemocytes engulf and digest the pathogens. They are equally important for encapsulation, where larger pathogens, such as parasitic larvae, are attacked by several hemocytes to isolate and neutralize them. Additionally, invertebrates make use of cellular and humoral responses, but also of inflammatory responses, when combating infections. Such responses include the induction of signaling pathways that result in the synthesis of toxic or injurious molecules for pathogens, such as reactive oxygen species. For instance, the phenoloxidase system in arthropods prepares reactive intermediates of phenolic compounds, which can deactivate pathogens (Cerenius & Söderhäll, 2021). Invertebrates reproduce so rapidly that they can maintain effective defenses even before a pathogen becomes established. Since more than 95% of all animal species are invertebrates, they have had millions of years to fine-tune their immune systems, evolving diverse mechanisms to provide immediate and broad-spectrum protection against pathogens. Unlike vertebrates, invertebrates do not have adaptive immunity with the vertebrate-acquired immunological memory of specific antigens. Nevertheless, unlike vertebrates, invertebrates lack such infectious events as an adaptive response against more complex pathogens and depend entirely on innate immunity as their first and most important line of defense. This innate immune system, though less versatile than the adaptive immune response, is highly effective at pathogen detection and elimination in real-time, without the need for prior exposure or memory (Janik-Karpinska et al., 2022). Recent research has highlighted that some invertebrates possess immune responses that exhibit features akin to adaptive immunity, including immune priming, where previous exposure to a pathogen enhances the immune response on subsequent encounters (Little & Kraaijeveld, 2004; Sheehan et al., 2020). This finding challenges the traditional view that invertebrates are restricted to innate immune responses and suggests the presence of evolutionary bridges between innate and adaptive immunity (Schulenburg et al., 2004).

Invertebrates use many innate immune strategies for protection against pathogens in multiple habitats on their journey to survival. These include pattern recognition, production of antimicrobial peptides (AMPs), cellular responses, and tissue repair mechanisms. The first line of defense comprises physical barriers such as exoskeletons, cuticles, and mucous membranes that prevent pathogen entry (Coates et al., 2022). However, when these barriers are breached, invertebrates rely on pattern recognition receptors (PRRs) to detect pathogen-associated molecular patterns (PAMPs) and initiate an immune response (Betancourt et al., 2024). These PRRs can recognize conserved structures on pathogens but not on the host. Upon binding to PAMPs, the PRRs activate intracellular signaling pathways that trigger immune responses (Macho & Zipfel, 2014). One of the most well-studied PRRs in invertebrates is the Toll receptor, first identified in Drosophila melanogaster (fruit fly), which is responsible for detecting bacterial and fungal components and activating an immune response (Mahanta et al., 2023). This discovery revolutionized our understanding of immune function, as it demonstrated the evolutionary conservation of the Toll-like receptors (TLRs) found in vertebrates.

In addition to the Toll receptors, invertebrates also utilize other pattern recognition receptors, such as peptidoglycan recognition proteins (PGRPs) and C-type lectins. These PRRs are crucial for the recognition of various pathogens, including bacteria and fungi, and play a pivotal role in activating the immune system in invertebrates (Wang et al., 2019b; Wang & Wang, 2013). Once a pathogen is recognized by these receptors, several intracellular signaling pathways are triggered that result in the production of antimicrobial peptides (AMPs). Known as AMPs, small proteins display a broad antimicrobial activity against a wide variety of pathogens, including bacteria, fungi, and viruses. Well-known examples of AMPs in invertebrates include defensins, cecropins, and attacins (Table 1.4), all of which are crucial for pathogen neutralization (Tassanakajon et al., 2015). These AMPs are important in invertebrate immune humoral arm and are quickly synthesized during infection. Central to the invertebrate immune defense against diverse pathogens is their action,

Table 1.4 Immune effector molecules in invertebrate humoral response

Molecule	Organism	Function	Pathogen targeted	Reference
Defensin	Drosophila	Membrane disruption	Gram+ bacteria	Hanson and Lemaitre (2020)
Cecropin	Mosquito	Pore formation	Broad-spectrum	Tassanakajon et al. (2015)
Attacin	Drosophila	Membrane disruption	Gram- bacteria	Hanson and Lemaitre (2020)
Myticin	Mussel	Antiviral activity	RNA viruses	Rodrigues et al. (2025)
Theromyzin	Crab	Lytic activity	Vibrio	Mahanta et al. (2023)
Phenoloxidase	Shrimp	Melanization	Bacteria/fungi	Coates et al. (2022)
Galectin	Earthworm	Agglutination, opsonization	Various	Cerenius and Söderhäll (2021)
Lectin	Oyster	Pathogen binding	Bacteria	Jia et al. (2016)
Peroxinectin	Crayfish	Phagocytosis mediation	Bacteria	Dong et al. (2011)
Arenicin	Polychaete worm	Lysis	Broad-spectrum	Orlov et al. (2019)

through disruption of the microbial cell membrane, thereby neutralizing or eliminating pathogens.

In addition to humoral immune responses, cellular immunity in invertebrates is mediated by hemocytes, which are functionally analogous to macrophages in vertebrates (Gupta, 1991). Hemocytes, the invertebrate version of blood cells, circulate in hemolymph, and draw in phagocytosis by taking in bacteria and digesting them. Phagocytosis is one of the most important cellular immune responses, enabling invertebrates to directly eliminate pathogens from their bodies (Sacchi et al., 2024). Encapsulation, especially of large pathogens such as parasitic larvae, which cannot be phagocytized by a single hemocyte, is also an important role for hemocytes. During encapsulation, multiple hemocytes surround the pathogen to isolate and neutralize it, preventing the pathogen from spreading (Eleftherianos et al., 2021).

Moreover, as a component of immune defense, inflammatory responses are also used by invertebrates, as by vertebrates. These responses involve the activation of immune signaling pathways, which lead to the synthesis of toxic molecules like reactive oxygen species (ROS) and other reactive intermediates that help to neutralize pathogens. In arthropods, the phenoloxidase system plays a critical role in generating phenolic compounds that are toxic to pathogens (Cerenius & Söderhäll, 2021). These molecules enable effective defenses against infections and an effective mechanism to clear out pathogens. However, invertebrates have developed highly efficient immune systems to detect, neutralize, and eliminate pathogens rapidly. Their innate immune systems have evolved to protect them from a wide variety of pathogens, and despite lacking adaptive immunity, they have developed sophisticated mechanisms to maintain survival in pathogen-rich environments (Guo et al.,

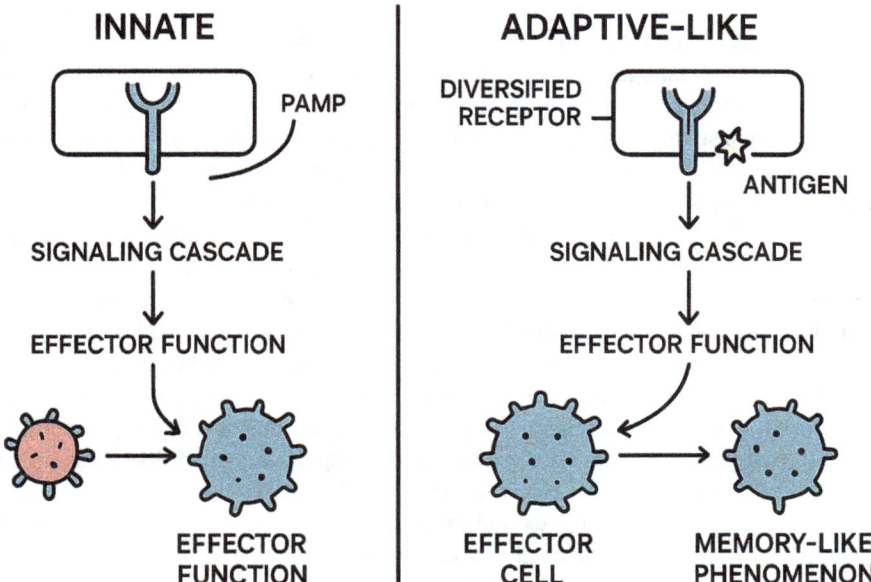

Fig. 1.4 Comparison and contrast of innate immune mechanisms with adaptive-like responses in invertebrates; signaling cascades, effector functions, and memory-like phenomena that blur the traditional division between innate and adaptive immunity in these organisms

2015). Conservation of immune responses across both invertebrates and vertebrates reveals similarities among immune function in diverse species. This conservation makes invertebrates invaluable models for understanding immune function and provides significant insights into the origins of immune systems (Buchmann, 2014). As our understanding of invertebrate immunity continues to emerge, intriguing possibilities exist for the use in the development of novel therapeutic strategies of interest in the management of disease, for pest control and antimicrobial therapy. And as we delve ever deeper into the immune systems of invertebrates, we will certainly discover even more elaborate immune pathways and mechanisms of broad importance to health of humans, agriculture, and biotechnology (Fig. 1.4). Furthermore, invertebrates offer a promising platform for the development of new antimicrobial agents to combat the growing threat of antimicrobial resistance pathogen (Destoumieux-Garzón et al., 2016; Guryanova et al., 2023).

1.7 Adaptive-Like Immunity in Invertebrates

Although several studies have suggested that at least some invertebrates have an immune response that is very much like adaptive immunity, the invertebrates do not have adaptive immunity in the classical sense. Mechanisms like clonal expansion, immune memory, and, in particular, pathogen recognition, which are features of adaptive immunity, are required for such responses. Some of these responses, including immune memory, were last shown to have taken place following bacterial infection in *D. melanogaster*. After re-exposure to the same pathogen, *Drosophila* appeared to mount a memory-type response where their immune system had a stronger response than what they mounted for the first time, but without having the induction of antibodies involved (Radtke et al., 2010). This was probably "immune memory" via epigenetic change, which will make the pathways easier to activate when they're infected again. Another example is another type of adaptive immunity in the *C. gigas*. An important characteristic of adaptive immunity is the ability of this mollusk to have all the machinery of a complex immune system, with recognition of pathogen-specific antigens. The oyster's immune system enables the organism to mount more specific immune responses to specific pathogens, especially bacteria and parasites (Allam & Raftos, 2015) through a series of immune receptors it produces, which are rather diverse. Yet, the reason for this enhanced specificity is likely the evolution of a more diversified immune repertoire, an ability to recognize and fend off more pathogens. Some studies have also shown that other invertebrate species, such as other arthropods and annelids, are capable of mounting adaptive immune-like responses. For example, earthworms (*L. terrestris*) are reported to exhibit immune priming (Ng et al., 2014), as exposure to pathogens induces an increase in immune response during further exposure. If true, that response has not reached the level of sophistication of the adaptive immune system of vertebrates, but it would mean immune memory developed far earlier among animals than had been believed (Fig. 1.5).

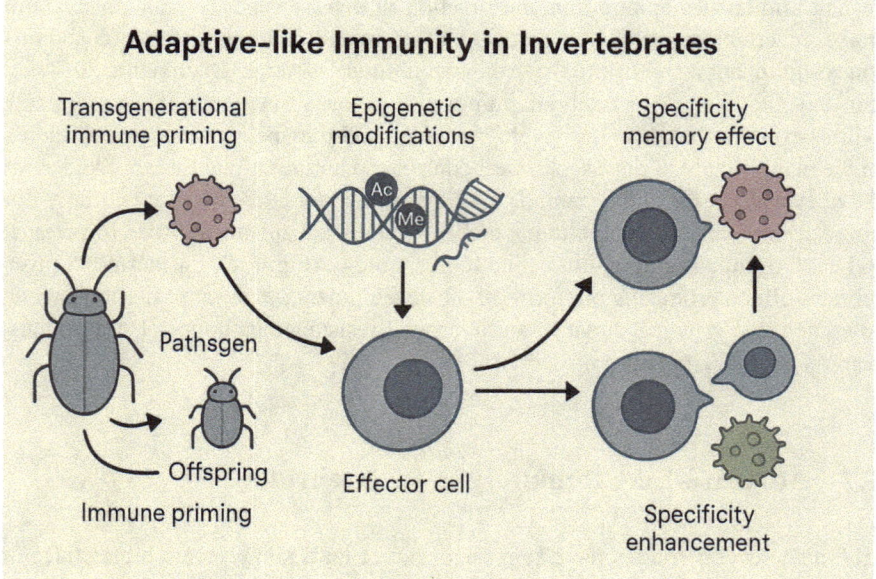

Fig. 1.5 Molecular and cellular processes underlying immune priming and memory-like functions in invertebrates. Includes examples of transgenerational immune priming and specificity enhancements through epigenetic and molecular modulations

1.8 Invertebrate Immune Mechanisms Across Phyla

Evolutionary adaptations and ecological niches of the nine major invertebrate phyla are related to their immune mechanisms. The Toll pathway is important in detecting bacteria and initiating an immune response in *D. melanogaster*. Mosquitoes, like other arthropods, use innate immune pathways to prevent the host invasion of pathogens such as *Plasmodium* and dengue viruses. The same mosquito research shows immune responses can impact disease transmission and helps understanding of usage of these responses in vector control (Baxter et al., 2017; Perveen et al., 2023). Humoral as well as cellular immunity are both important in mollusks such as the Pacific oyster, *Crassostrea gigas*. These organisms do phagocytosis through hemocytes, and they produce several AMPs to ensure their protection from microbial infection. For example, the immune receptors the oysters have to detect the pathogens' molecules are an adaptive-like mechanism, and therefore the oysters will respond more specifically to the pathogens' molecules (Wang et al., 2019a, b). Such annelids, e.g., *Lumbricus terrestris,* have cellular immunity in which the coelomic fluid contains phagocytic cells that engulf and neutralize the pathogens. Immune priming in earthworms, developed like in other invertebrates, is also typical, i.e., it involves an increase in the immune response against a pathogen after the second encounter with the antigen (Engelmann et al., 2011). Echinoderms also develop humoral immune responses but are similar to echinoderms (starfish, sea urchins) in having components from cells. When echinoderms, like other invertebrates, become infected, they produce a variety of AMPs, and, like other invertebrates, immune cells (coelomocytes) engulf pathogens (Coates et al., 2022).

1.9 Evolutionary Significance and Future Directions

Important insights into the evolution of immune systems across the animal kingdom are derived from the study of invertebrate immune responses. Invertebrates, which rely exclusively on innate immunity, have evolved pattern recognition receptors (PRRs), antimicrobial peptides (AMPs), and cellular immune mechanisms that provide a rapid, nonspecific response to pathogens (Patnaik et al., 2024). These immune components form the basis upon which the more complex vertebrate immune systems were founded, suggesting that vertebrates inherited and modified these early immune mechanisms (Boehm et al., 2012). The identification of adaptive-like immune responses in certain invertebrates, such as immune priming—where prior exposure to pathogens results in a stronger immune response upon subsequent exposure—has raised intriguing questions about the evolution of immune specificity and the origins of immune memory (Cooper & Eleftherianos, 2017). While traditionally thought to lack the ability for immune memory, some invertebrates exhibit enhanced responses following re-exposure to pathogens, indicating an early form of immune memory (Gourbal et al., 2018). These findings suggest that there may be further instances of adaptive-like immunity in invertebrates yet to be discovered, which could offer significant insight into the evolutionary adaptation of immune systems.

Advances in genomic, transcriptomic, and proteomic technologies, such as CRISPR-Cas9 and RNA sequencing, have propelled the study of invertebrate immunity to new heights, allowing for the identification of previously unknown immune pathways and genes (Sindelar, 2024; Söderhäll, 2024). High-throughput RNA sequencing (RNA-Seq), for example, has enabled large-scale analyses of immune gene expression in invertebrates, providing a comprehensive understanding of immune signaling and pathogen recognition mechanisms (Clark & Greenwood, 2016). The sequencing of genomes in species like *D. melanogaster* and *C. gigas* has provided valuable insights into immune gene families, including those related to Toll-like receptor (TLR) pathways and IMD signaling (Lemaitre & Hoffmann, 2007; Sekino et al., 2012), reinforcing the evolutionary relationship between invertebrates and vertebrates. These technological advancements are unlocking new avenues for understanding immune responses at the molecular level and are helping researchers discover novel immune effectors and antimicrobial agents (Zhang et al., 2011).

The identification of adaptive-like immunity in invertebrates is one of the most fascinating discoveries of recent years, challenging the longstanding view that these organisms only possess innate immunity. While invertebrates lack the clonal selection and antigen-specific memory seen in vertebrates, their ability to exhibit immune priming suggests a mechanism of immune response that may offer a bridge between innate and adaptive immunity (Wang et al., 2025). This has significant implications for the evolution of immune memory and opens up new avenues for research into the development of immune specificity in early life forms. In addition to its evolutionary significance, the study of invertebrate immunity has important practical applications in fields such as biotechnology, pest control, and disease management. In aquaculture, for example, immunostimulants derived from invertebrate immune pathways have shown promise in enhancing disease resistance in species like shrimp, oysters, and mussels (Kumar et al., 2023; Rodrigues et al., 2025). The discovery of antimicrobial peptides (AMPs) in invertebrates has also led to the development of novel antimicrobial agents that could serve as alternatives to traditional antibiotics, addressing the growing threat of antimicrobial resistance (Lazzaro et al., 2020). Additionally, immunotherapies and vaccines based on invertebrate immune systems are being explored to boost the immune responses of both aquatic species and terrestrial crops (Woods et al., 2017). As research continues to evolve, the integration of advanced technologies like genomics and bioinformatics will undoubtedly uncover novel immune mechanisms and regulatory pathways. This will not only deepen our understanding of invertebrate immunity but also provide innovative solutions in disease control, pest management, and the treatment of emerging infectious diseases. The field of invertebrate immunology offers great promise for developing sustainable and effective therapies in both human health and agriculture.

1.10   Cellular Immunity in Invertebrates

1.10.1   Hemocytes: Types and Functions

The primary cellular effectors of invertebrate immunity are the hemocytes (invertebrate white blood cells). (Amaral et al., 2010) describe that they are a part of phagocytosis, encapsulation, and nodule formation. In different invertebrate taxa, several types of hemocytes have been identified including prohemocytes, granulocytes, plasmatocytes, and oenocytoids with specific immune roles (Eleftherianos et al., 2021). The most abundant granulocytes and plasmotocytes are usually involved in pathogen recognition and clearance (Greenlee-Wacker, 2016). For instance, plasmatocytes in insects are mainly engaged in phagocytosis and encapsulation, while in crustaceans, granular cells liberate antimicrobial peptides and enzymes contained in cytoplasmic granules (Lavine & Strand, 2002) (Fig. 1.6).

1.10.2   Phagocytosis and Encapsulation

A fundamental process in innate immunity is the engulfment and digestion of invading pathogens by specialized immune cells, a mechanism known as phagocytosis. This cellular defense strategy enables organisms to directly eliminate microbes,

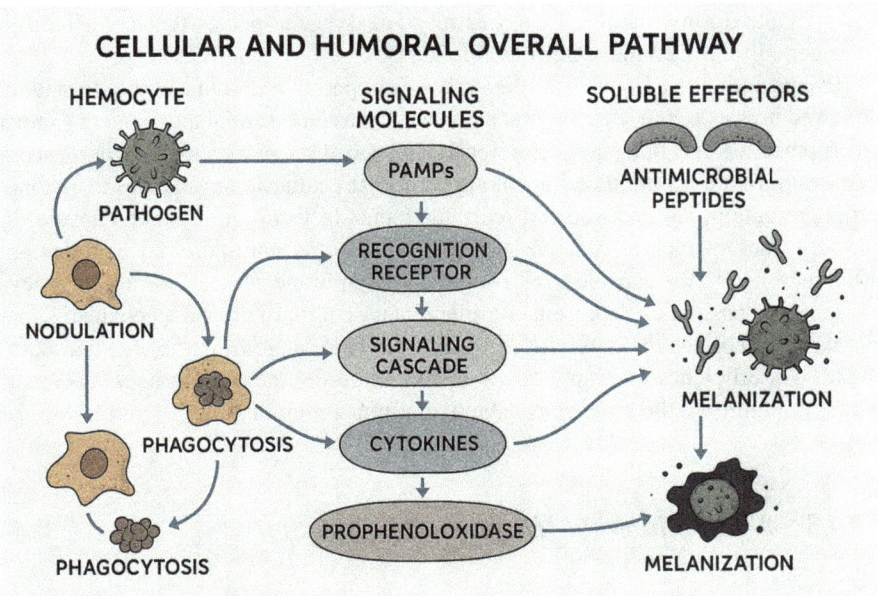

Fig. 1.6 Integration of cellular and humoral immune responses into a unified schematic. It shows interactions between hemocytes, signaling molecules, and soluble effectors coordinating a comprehensive defense strategy

preventing their proliferation and dissemination within host tissues. The initiation of phagocytosis begins with the recognition of pathogens through pattern recognition receptors (PRRs) expressed on the surface of phagocytic cells. PRRs identify conserved microbial structures known as pathogen-associated molecular patterns (PAMPs), including components such as lipopolysaccharides (LPS) from Gram-negative bacteria, β-glucans found in fungal cell walls, and peptidoglycans typical of bacterial envelopes (Silva-Gomes et al., 2014). Upon binding PAMPs, PRRs trigger intracellular signaling cascades that activate and coordinate the immune response. In invertebrates such as *Drosophila melanogaster*, key pathways involved include the Toll, immune deficiency (Imd), and Janus kinase/signal transducer and activator of transcription (JAK/STAT) pathways (Myllymäki & Rämet, 2014). These signaling networks induce the expression of antimicrobial peptides (AMPs) and other immune effectors that enhance the host's ability to combat infection. In addition to promoting pathogen clearance, these pathways regulate phagocytic activity by modulating the expression of receptors and cytoskeletal components essential for engulfment. While phagocytosis effectively controls many microbial pathogens, larger invaders such as parasitic helminths or parasitoid eggs cannot be ingested due to their size. To counteract these, invertebrates have evolved an alternative defense known as encapsulation. This process involves the recruitment and aggregation of hemocytes—circulating immune cells—that adhere to the surface of the large foreign body and form a multilayered cellular capsule around it (Melillo et al., 2018). The capsule often undergoes melanization, a biochemical reaction mediated by the phenoloxidase cascade, resulting in the deposition of melanin around the invader. Melanin not only physically isolates the pathogen but also generates cytotoxic intermediates such as reactive oxygen species that contribute to pathogen killing (Cerenius & Söderhäll, 2021).

Encapsulation is a highly coordinated and cooperative immune response widely observed in insects and some crustaceans, serving as an essential mechanism against macroparasites and other large foreign bodies (Rowley et al., 2022). The process requires intricate communication among hemocytes, adhesion molecules to mediate cell aggregation, and cytoskeletal rearrangements to form an effective capsule. In addition, factors such as reactive nitrogen species and antimicrobial peptides are often recruited to the capsule site to augment the immune attack. Overall, phagocytosis and encapsulation represent complementary cellular defense mechanisms that enable invertebrates to respond effectively to a wide range of pathogens and parasites. Their efficiency and rapid deployment underscore the evolutionary success of innate immunity in the absence of adaptive immune memory.

1.10.3 Nodulation and Hemocyte Aggregation

Besides phagocytosis and encapsulation, invertebrates have another important immune response for preventing microbial infection, nodulation. Nodulation involves the aggregation of hemocytes (the equivalent of macrophages in vertebrates) around clusters of microbes to form nodules, which are subsequently melanized and killed. This process is particularly important for the defense against bacterial infections and other pathogens that are too large for a single hemocyte to handle through phagocytosis alone (Liu et al., 2020). Hemocytes surround a pathogen during nodulation eventually isolating the pathogen within a nodule. The nodule is then melanized, a process in which the pathogen is coated with melanin, effectively encapsulating and killing the pathogen (Liu et al., 2020). This mechanism is vital for invertebrates, as it provides a way to neutralize large or complex pathogens such as parasitic larvae or fungal spores, which may evade direct phagocytosis. Nodulation is a rapid immune response that often precedes encapsulation, which is the formation of a protective barrier around the pathogen by multiple hemocytes (Satyavathi et al., 2014). It appears almost instantaneously after a pathogen is detected, usually minutes to hours after a pathogen infects the tissue. The process of nodulation is bacterial cell wall-dependent; it is activated by conserved components of the pathogen's cell wall, such as lipopolysaccharides (LPS) in Gram-negative bacteria, which serve as potent pathogen-associated molecular patterns (PAMPs) recognized by the host's immune receptors (Schaefer et al., 2018). Cytokine-like molecules play a significant role in regulating the nodulation process. The role of these molecules in their participation in activating the immune response and the coordination of hemocyte activities is implicated here. Similar to cytokines in vertebrates, these immune modulators help facilitate communication between immune cells, ensuring a coordinated and effective immune response (Salvador et al., 2021). Initiation and maintenance of the hemocyte aggregation around the pathogen and regulation of nodule melaninization for ultimate pathogen neutralization depend on release of these molecules. The melanization in the nodule is another important step in this immune defense. Not only is melanin a physical barrier, but it also becomes a toxic agent that directly reduces viability of the pathogen. This is especially important in invertebrates, as melanin production is cytotoxic and can destroy pathogens by generating highly reactive free radicals (Nappi & Ottaviani, 2000). Furthermore, melanization is coupled with the formation of reactive oxygen species (ROS), which are essential in killing pathogens. These ROS play a critical role in oxidative killing of the microbes within the nodule, ensuring that the pathogen is effectively neutralized. Recent studies have shown that nodulation is particularly important in arthropods, such as insects and crustaceans, which rely on this rapid immune response as part of their innate immunity (Cerenius & Söderhäll, 2021). For these organisms, nodulation is an early defense with immediate protection against a broad range of pathogens from microorganisms such as bacteria, fungi, and parasites. In some cases, nodulation can occur before more complex immune responses, such as encapsulation, are initiated, highlighting the dynamic and layered nature of invertebrate immune defense (Satyavathi et al., 2014).

The nodulation process in invertebrates is, however, important but remains poorly understood, and research reveals new immune receptors, signaling pathways, and cytokine-like molecules that regulate this immune response. Molecular techniques such as genomics and proteomics are advancing rapidly, and these advances are also beginning to provide us with invaluable insights into the genetic and molecular basis of nodulation. For example, recent studies have identified novel PRRs and signaling molecules involved in nodulation in *D. melanogaster*, providing new targets for immune modulation in disease management (Li & Chang, 2021; Mahanta et al., 2023). Additionally, studies on the genetic regulation of nodulation in crustaceans have shown that the process can be modulated by both innate immune genes and environmental factors, further demonstrating the plasticity of the invertebrate immune system. Finally, we conclude that nodulation is a key cellular immune response of invertebrates, largely responsible for innate immune defense against invertebrate pathogens, which cannot be simply avoided by phagocytosis. Encapsulation and melanization, as well as this process, constitute an efficient and rapid, yet broad spectrum, defense mechanism utilized by many invertebrates to survive in pathogen-rich environments. These results contribute to a complex and dynamic picture of invertebrate immunity that continues to inform our understanding of the evolutionary roots of immune responses among animals.

1.11   Humoral Immunity in Invertebrates

1.11.1   Antimicrobial Peptides (AMPs)

Antimicrobial peptides (AMPs) are one of the most potent and evolutionarily conserved weapons of the humoral immunity in invertebrates. The small cationic peptides have broad spectrum activity against bacteria, fungi, and some viruses (Rodrigues et al., 2025; Tassanakajon et al., 2015). *Drosophila melanogaster* have been critical to AMP research and have revealed some key molecules that are produced mainly in the fat body and hemocytes, in response to microbial challenge (Hanson & Lemaitre, 2020). Expression of AMPs is usually induced through the Toll and Imd signaling pathways upon recognition of pathogen-associated molecular patterns (PAMPs) by peptidoglycan recognition proteins (PGRPs) and Gram-negative binding proteins (GNBPs) (Salcedo-Porras et al., 2021). NF-κB transcription factors control their transcription as drivers of a strong and immune specific response (Vallabhapurapu & Karin, 2009). AMPs like penaeidins, crustins, and mytilins are shown to be important in defense against aquatic pathogens in crustaceans and mollusks (Rosa & Barracco, 2010). In recent studies, (Clark et al., 2020) present the evidence that AMP diversity in mollusks is far more extensive than believed before, with lineage specific expansion that represents the adaptive evolutionary pressure of aquatic environment.

1.11.2   Phenoloxidase and the Prophenoloxidase Cascade

In invertebrates, melanization is a vital immune process protecting against pathogens, as encapsulated and killed pathogens represent dead ends to pathogen spread. This process is mediated by the enzyme phenoloxidase (PO), a copper-containing enzyme that plays a central role in the immune response of many invertebrates, including arthropods and crustaceans. PO is initially present in an inactive precursor form known as prophenoloxidase (proPO), which must be activated to become functional. The activation of proPO occurs through a proteolytic cascade initiated by the recognition of pathogens (Jearaphunt et al., 2014). During serine protease recognition of microbes by the immune system, proPO is cleaved into its active form, PO. The activated PO then functions as an oxidase to oxidize phenolic compounds into quinones, leading to the production of melanin at the site of infection or injury (Kumar et al., 2024). Multiple functions of melanization are pathogen defense. A key function is pro insofar as they trap and entrap, thereby encapsulate pathogens—effectively isolating the invader from the host tissue and its spread. Moreover, the oxidation of phenols generates toxic intermediates, including quinones and other reactive molecules, that kill the trapped pathogens by disrupting their cellular structures and functions (Ito et al., 2020). These toxic intermediates are particularly effective against bacteria, fungi, and some parasites, making melanization an essential part of the invertebrate immune system. Melanization is also chemically neutralizing in this way and physically isolating the pathogen. However, melanization must be tightly regulated to avoid its overpolymerization or misactivation, leading to damage of the host's own tissues. Serine protease inhibitors (serpins) play a key role in regulating the proPO system. The activation of proPO is controlled (by these inhibitors) so that melanization will take place only as required. By preventing premature activation of proPO or overactivation during immune responses, serpins prevent unnecessary tissue damage and ensure that the immune response is localized to the site of infection (Gatto et al., 2013). This regulatory mechanism ensures that the host can fight infection while not risking collateral damage from the very response that enables fighting infection.

ProPO provides an attractive system for studying these reactions because it has been so well studied in various invertebrates, including crustaceans such as shrimps and crayfish, where the system is important for antibacterial and antifungal defense. Melanization is absolutely required for these organisms to mount an immune response that is effective against microbial threats. For example, the proPO system is important for defense against bacterial and fungal infection in shrimp. Knockdown or inhibition of proPO activity significantly impairs the immune response, leaving the shrimp more vulnerable to pathogens (Amparyup et al., 2013). The proPO system is critical for survival in crustaceans and other arthropods and, despite the evolutionary divergence of the active component (phenol oxidases), shares many similarities with insect immunity, indicating its significance to invertebrate as a whole. The role of melanization and the proPO system in pathogen defense in insects is well known, but it is additionally involved in wound healing in invertebrates. Melanin deposition around wounds does more than encapsulate invading pathogens but also seals the wound to stop further infection. In environments such as predator/predated and environmental stressors where physical injury is common, maintenance of tissue integrity requires a rapid response.

Melanization is a very important immune response in invertebrates, which, taken as a whole, serves as physical and chemical barriers to infection. Actuation of the proPO system allows invertebrates to mount a rapid, effective response to an extensive array of pathogens. This process is tightly regulated by serpins in order to use it appropriately, killing pathogens as desired, but reducing the risk of tissue damage from it. Given that melanization plays a central role in invertebrate immune defense, the study of this process is informative with regard to immune defense mechanisms and presents considerable potential for use in biotechnology, as a means of pest control and for the development of therapeutics. The molecular mechanisms of melanization and the proPO system may also yield novel avenues for development of antimicrobial agents directed at these immune pathways.

1.11.3   Lysozymes and Reactive Oxygen Species

The enzymatic components of humoral immunity, namely lysozymes, hydrolyze the β-(1,4)-glycosidic bonds in the bacterial peptidoglycan, primarily targeting the Gram-positive bacteria (Ferraboschi et al., 2021; Ragland & Criss, 2017). They are present in hemolymph and in gut and hemocytes and may cooperate with AMPs to ensure effective immune protection (Zhang et al., 2024). Similar to that in vertebrates, production of reactive oxygen species (ROS) such as superoxide anions and hydrogen peroxide during the respiratory burst also takes place in invertebrates (Chainy et al., 2016). Microbial killing and signaling use these molecules, but because they must be tightly regulated to avoid host tissue damage, they are large molecules.

1.11.4   Lectins and Agglutinins

Carbohydrate-binding proteins (lectins) that function as nonself-recognition molecules and as mediators of immune responses are common in invertebrates (Vasta et al., 2007). Binding to specific sugar moieties on pathogen surfaces is carried out by them and facilitates agglutination, opsonization, and activation of downstream immune cascades. Among invertebrates, particularly well-studied are the mannose-binding lectins and C-type lectins in shrimp, mollusks, and echinoderms (Ahmmed et al., 2022). In fact, they often act as opsonins that help enhance phagocytosis or stimulate the proPO cascade (Cerenius et al., 2008). The processes mentioned are also occupied by ficolins and galectins serving as critical mediators in innate immunity (Endo et al., 2015; Liu & Stowell, 2023).

1.12   Signaling Pathways and Immune Regulation

1.12.1   *Toll and Imd Pathways*

In many invertebrate species, such as insects, Drosophila melanogaster, for example, two main signaling cascades govern immediate immunity: Toll and Imd (immune deficiency) pathways (Cammarata-Mouchtouris et al., 2022). The Gram-positive bacteria and fungi trigger the Toll pathway, which is present in most tissues, while the Gram-negative bacteria use the Imd pathway, which is primarily found in the fat body (Alejandro et al., 2022). Upon the recognition of pathogen, PRRs like PGRPs and GNBPs start a signaling cascade bringing about the activation of Dorsal and Dif NF-κB transcription factors in the Toll pathway and Relish in the Imd pathway (Cammarata-Mouchtouris et al., 2022). These factors can translocate to the nucleus and cause the expression of AMPs and other immune effectors. Further evidence also pointed out that the Toll pathway may play a role in antiviral responses as well as developmental regulation (Zambon et al., 2005) (Fig. 1.7).

1.12.2   *JAK/STAT and JNK Pathways*

Hemocyte proliferation, systemic immune response, and antiviral defense are all dependent on the Janus kinase/signal transducer and activator of transcription (JAK/STAT) pathway in insects and crustaceans (Liu et al., 2009). The nuclear protein can be activated by cytokine-like molecules and regulates genes associated to stress responses, inflammation, and immunity (Yamamoto & Takeda, 2008). Likewise, the

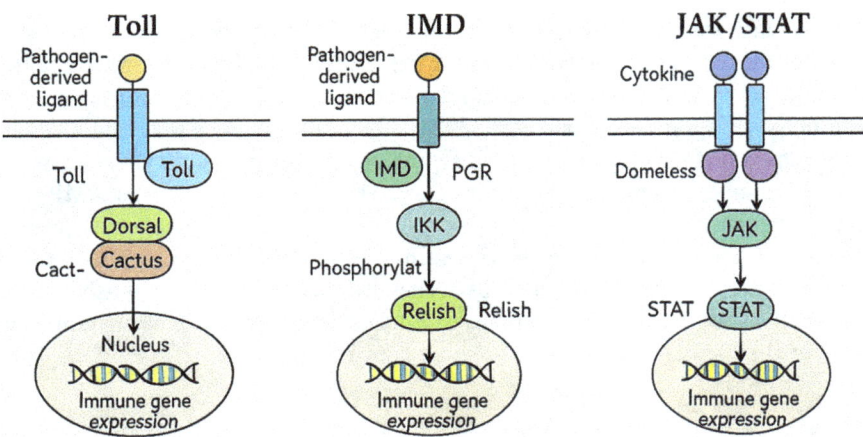

Fig. 1.7 Mapping of key intracellular signaling cascades (e.g., Toll, IMD, JAK/STAT) activated upon pathogen recognition in invertebrate immune cells. It also describes ligand-receptor interactions, secondary messengers, and transcription factors involved in immune gene expression

Jun N terminal kinase (JNK) pathway activates immune genes, apoptosis, and wound healing. In parallel with the Imd pathway, it acts and contributes both to epithelial defense and stress signaling (Zhai et al., 2018). Due to the very complex and integrated nature of invertebrate immune system, these signaling pathways often intersect or cross regulate.

1.12.3   Negative Regulation and Immune Homeostasis

While immune activation is necessary for the defense against pathogen infection, it or its association with chronic inflammation, can cause excessive or chronic immune responses, which are detrimental to the host. Thus, negative regulators of immune signaling are important to maintain homeostasis (Soares et al., 2017). The few negative regulators of Imd and Toll signaling pathways were identified in Drosophila, including Pirk, Caspar, and Dnr1 pathway signaling inhibition (Cammarata-Mouchtouris et al., 2022; Kleino et al., 2008). Feedback inhibition and serpins acting as protease inhibitors regulate the immune response and prevent self-destruction. Similar regulatory mechanisms are seen in aquatic invertebrates. For instance, in crustacean studies, the abundance of proPO activation can be harmful (Amparyup et al., 2013), and serpins are there to control this response.

1.13   Immune Memory and Priming in Invertebrates

1.13.1   Evidence for Immune Priming

Invertebrates are traditionally thought incapable of mounting adaptive immunity displays but have been shown to possess the ability to be immune primed, which refers to previously being exposed to a pathogen that increases protection upon being exposed again (Rathinam et al., 2024). The priming can be specific for certain pathogens and last for an organism's lifetime or across generations (Dhinaut et al., 2018). Immune priming also increases survival and the expression of AMPs and phagocytic activity after a secondary exposure in experimental studies on invertebrates like *Tribolium castaneum* (Prakash et al., 2024) and Tenebrio molitor. Likewise, exposure to bacteria in crustaceans such as shrimp and crayfish leads to enhancement of immune response and reduced pathogen load when reinfected (Tran et al., 2022).

1.13.2   Mechanisms of Immune Memory

There are still many questions about how immune memory arises in invertebrates. Some are the long–term activation of immune pathways, epigenetics, and hemocyte population changes (Melillo et al., 2018). In contrast to vertebrate memory mediated by lymphocytes, how invertebrate immune memory works is more likely due to enhanced readiness of innate immune cells and tissues. More recently, other pattern recognition receptors (PRR) with a memory-like behavior have been suggested in the immune response of mollusks, such as the fibrinogen-related protein (FREP), and in insects such as down syndrome cell adhesion molecule (Dscam), which could provide specificity and diversity (Li et al., 2025). The extent of Dscam diversity from alternative splicing could be involved in immune specificity in crustaceans as in antibodies (Ng & Kurtz, 2020). Aside from this, maternal immune priming is found in many species, where immune factors are transferred from mother to offspring, which helps the offspring in resisting the infections.

1.13.3   Implications for Disease Management

Immune priming in invertebrates is of profound importance for the aquaculture, pest control, and ecological management. While several strategies, like using immunostimulants or sublethally exposing invertebrate hosts to a pathogen may prime the hosts and enhance invertebrate resistance to disease (Best et al., 2013), these methods may also inadvertently introduce pathogens to the entire batch of organisms involved. Immune priming in natural populations may play a role in host pathogen coevolution and play a role in the dynamics of population (Dubief et al., 2017). These insights overturn the other dichotomy between innate and adaptive immunity of invertebrate immunity and demonstrate the plasticity of the invertebrate immune response.

1.14   Applications and Future Perspectives

1.14.1   Biomedical and Biotechnological Applications

Biomedical science and biotechnology work by bringing a broad range of approaches to the generation of novel therapeutic agents and biotech solutions are enabled, in part, by the contributions of the invertebrate immune system. However, the rich diversity of immune components in invertebrates has provided a bonanza of bioactive molecules of potential use in medicinal and industrial areas. One of the most noteworthy of these contributions is the finding and use of antimicrobial peptides (AMPs) produced by various invertebrates. Broad spectrum activity of these peptides as potent antimicrobial agents against bacteria, fungi, and viruses. AMPs have a low potential to induce resistance and are therefore attractive alternatives to mitigate the problem of AMR (Mba & Nweze, 2022). AMPs derived from invertebrates such as crustaceans, insects, and mollusks are very attractive as new therapeutic agents, largely responsible to AMPs being used in topical creams for wound healing and systemic therapy for infection (Pfalzgraff et al., 2018). In addition to AMPs, hemocyanins—oxygen-carrying protein found in crustacean blood—have also been eyed as immunostimulants and adjuvants in vaccine development. The discoveries reached to date show that hemocyanins can actually serve as a source of immune activators, stimulations of which can significantly enhance the body's immune response helpful for counteracting infections (Zhong et al., 2016). Like lectins, a class of carbohydrate-binding proteins have been shown to possess promise as immunomodulators in therapeutic formulations because they can recognize and bind carbohydrates specific to a pathogen, thereby triggering immune responses (Nabi-Afjadi et al., 2022). Inspired by the structural diversity and functional versatility of invertebrate immune effectors, we have made significant advances in the development of biosensors, vaccines, and drug delivery systems using these immune components as biotechnological platforms for the development of novel immunotherapies, especially for cancer, vaccines, and biodegradable drug delivery systems. Biomolecular scaffolds that use invertebrate-derived proteins and molecules serve as sensitive biosensors that can detect a range of pathogens, environmental toxins, or even the presence of disease markers in humans (Bhatia et al., 2024). In addition, invertebrate models such as the fruit fly Drosophila melanogaster and the nematode Caenorhabditis elegans have particularly well-characterized immune systems that are conserved among other organisms and are commonly used in laboratory research. The power of these models is that they enable scientists to test the molecular mechanisms that govern immune responses, disease progression, and indeed, genetic regulation of immune function. Finally, they are invaluable also in understanding innate immunity-relevant diseases, including neurodegeneration, autoimmunity, and chronic inflammation (Cappellano et al., 2013). These models have allowed the use of genetic tools and high-throughput screening to lead to the discovery of new immune signaling pathways, as well as to which dynamic relationships can be utilized in the context of immune regulation for both invertebrate and vertebrate species.

1.14.2 Aquaculture

Immunity of invertebrates is of critical importance for disease management and health monitoring in aquaculture. Because of their aquatic environment, which frequently accommodates variety of pathogen, aquatic organisms such as shrimp, oysters, and mussels are highly susceptible to infections. Immune priming has been widely used in aquaculture practices to boost disease resistance in these important, economically significant species, and the mechanisms of invertebrate immunity are now well known. The term by which increased immune response to a pathogen is observed after a previous exposure to the same pathogen, similar to an adaptive immune response in vertebrates is known as immune priming (Riera Romo et al., 2016). Immune stimulants are commonly fed to invertebrates in aquaculture, putting the food in them and introducing them to more natural immune survival, helping to keep disease outbreaks at bay, and invertebrates in aquaculture reap the benefit of gene-based screening tools to pick the ones more resistant to disease, making healthier populations. Dietary supplementation of immunostimulants plus these techniques is now an integral part of health management systems used in the aquatic farming industry (Bricknell & Dalmo, 2005). One application is disease management in shrimp—for example, animal resistance to disease has been enhanced through use of beta-glucans, other immune-boosting compounds and products derived from marine invertebrates (Uengwetwanit et al., 2025). Bivalves and crustaceans are also widely used in environmental monitoring and pollution assessment. These organisms function as bioindicators, and the immune responses of these organisms serve as sensitive markers of heavy metals, microplastics, and chemical contaminant stressors. These organisms' immune biomarkers reflect pollutant exposure and are valuable indicators of the health of marine ecosystems. In particular, such organisms (e.g., mussels and oysters) have been found to have immune responses that correlate directly with the levels of contaminants in their environments, making them useful bioindicator organisms for assessing environmental pollution (Balbi et al., 2021). By these examples, we show that invertebrate immunity has a wider application than disease resistance; it is relevant to ecological monitoring and sustainable management.

1.14.3 Future Directions

In contrast, invertebrate immunity has developed significantly as a field, yet there are still several black holes of our ignorance of its complexity and regulation. Much future research must be conducted to further study functional genomics of immune genes for varying invertebrate species, specifically addressing the genetic basis of immune response and molecular crosstalk between the emerging immune system and other physiological systems such as nervous and endocrine systems. Another area deserving of closer attention is epigenetic regulation of immune response, which could help understand how environmental factors and genetic variants can affect immune system's response to pathogens. The promise of highly promising integrative omics technologies, such as transcriptomics, proteomics, and metabolomics, conferred with high-throughput sequencing approaches, toward revealing the novel immune effectors and regulatory mechanisms of one phylum on another is enormous. The application of these methods will almost certainly lead to new insights in the evolution of immune systems, uncovering unrecognized immune components or pathways.

In addition, knowing how immune complexity evolved and how immune memory arose in invertebrates will be key to this puzzle. The study of invertebrates can help provide a better understanding of the putative evolutionary history of vertebrate adaptive immunity and of how particular adaptive immune mechanisms may have contributed to the evolution of immunological memory. Useful information from this research will also help guide further translational studies such as vaccine development, immunotherapy, and antimicrobial development. Overall, the future of invertebrate immunology rests in how we can increase our knowledge base of immune gene function, evolutionary biology in general, and the interchange between immune systems and environmental stressors. As new biotechnological tools are discovered and research methodologies become more and more advanced, the invertebrate immune system is sure to reveal more and more important discoveries that will greatly impact biomedical research, aquaculture, and environmental health.

References

Ahmmed, M. K., Bhowmik, S., Giteru, S. G., Zilani, M. N. H., Adadi, P., Islam, S. S., Kanwugu, O. N., Haq, M., Ahmmed, F., Ng, C. C. W., Chan, Y. S., Asadujjaman, M., Chan, G. H. H., Naude, R., Bekhit, A. E.-D. A., Ng, T. B., & Wong, J. H. (2022). An update of lectins from marine organisms: Characterization, extraction methodology, and potential biofunctional applications. *Marine Drugs, 20*(7), Article 7. https://doi.org/10.3390/md20070430

Alejandro, A.-D., Lilia, J.-P., Jesús, M.-B., & Henry, R. M. (2022). The IMD and Toll canonical immune pathways of Triatoma pallidipennis are preferentially activated by gram-negative and gram-positive bacteria, respectively, but cross-activation also occurs. *Parasites & Vectors, 15*(1), 256. https://doi.org/10.1186/s13071-022-05363-y

Allam, B., & Raftos, D. (2015). Immune responses to infectious diseases in bivalves. *Journal of Invertebrate Pathology, 131*, 121–136. https://doi.org/10.1016/j.jip.2015.05.005

Amaral, I. M. R., Neto, J. F. M., Pereira, G. B., Franco, M. B., Beletti, M. E., Kerr, W. E., Bonetti, A. M., & Ueira-Vieira, C. (2010). Circulating hemocytes from larvae of *Melipona scutellaris* (Hymenoptera, Apidae, Meliponini): Cell types and their role in phagocytosis. *Micron, 41*(2), 123–129. https://doi.org/10.1016/j.micron.2009.10.003

Amparyup, P., Sutthangkul, J., Charoensapsri, W., & Tassanakajon, A. (2012). Pattern recognition protein binds to lipopolysaccharide and β-1, 3-glucan and activates shrimp prophenoloxidase system. *Journal of Biological Chemistry, 287*(13), 10060–10069.

Amparyup, P., Charoensapsri, W., & Tassanakajon, A. (2013). Prophenoloxidase system and its role in shrimp immune responses against major pathogens. *Fish & Shellfish Immunology, 34*(4), 990–1001. https://doi.org/10.1016/j.fsi.2012.08.019

Armitage, S. A., Sun, W., You, X., Kurtz, J., Schmucker, D., & Chen, W. (2014). Quantitative profiling of Drosophila melanogaster Dscam1 isoforms reveals no changes in splicing after bacterial exposure. *PLoS One, 9*(10), e108660.

Balbi, T., Auguste, M., Ciacci, C., & Canesi, L. (2021). Immunological responses of marine bivalves to contaminant exposure: Contribution of the -omics approach. *Frontiers in Immunology, 12*, 618726. https://doi.org/10.3389/fimmu.2021.618726

Baluška, F., Miller, W. B., & Reber, A. S. (2023). Cellular and evolutionary perspectives on organismal cognition: From unicellular to multicellular organisms. *Biological Journal of the Linnean Society, 139*(4), 503–513. https://doi.org/10.1093/biolinnean/blac005

Baxter, R. H. G., Contet, A., & Krueger, K. (2017). Arthropod innate immune systems and vector-borne diseases. *Biochemistry, 56*(7), 907–918. https://doi.org/10.1021/acs.biochem.6b00870

Best, A., Tidbury, H., White, A., & Boots, M. (2013). The evolutionary dynamics of within-generation immune priming in invertebrate hosts. *Journal of the Royal Society, Interface, 10*(80), 20120887. https://doi.org/10.1098/rsif.2012.0887

Betancourt, J. L., Rodríguez-Ramos, T., & Dixon, B. (2024). Pattern recognition receptors in Crustacea: Immunological roles under environmental stress. *Frontiers in Immunology, 15*, 1474512. https://doi.org/10.3389/fimmu.2024.1474512

Bhatia, D., Paul, S., Acharjee, T., & Ramachairy, S. S. (2024). Biosensors and their widespread impact on human health. *Sensors International, 5*, 100257. https://doi.org/10.1016/j.sintl.2023.100257

Bisola, M.-A. I., Olatunji, G., Kokori, E., Mustapha, A. A., Scott, G. Y., Ogieuh, I. J., Woldehana, N. A., Stanley, A. C., Olohita, O. A., Abiola, A. S., Olawade, D. B., & Aderinto, N. (2024). Emerging challenges in innate immunity: *Staphylococcus aureus* and healthcare-associated infection. *Journal of Medicine, Surgery, and Public Health, 3*, 100103. https://doi.org/10.1016/j.glmedi.2024.100103

Boehm, T., Iwanami, N., & Hess, I. (2012). Evolution of the immune system in the lower vertebrates. *Annual Review of Genomics and Human Genetics, 13*(Volume 13, 2012), 127–149. https://doi.org/10.1146/annurev-genom-090711-163747

Boraschi, D., Alijagic, A., Auguste, M., Barbero, F., Ferrari, E., Hernadi, S., Mayall, C., Michelini, S., Navarro Pacheco, N. I., Prinelli, A., Swart, E., Swartzwelter, B. J., Bastús, N. G., Canesi, L., Drobne, D., Duschl, A., Ewart, M.-A., Horejs-Hoeck, J., Italiani, P., et al. (2020). Addressing nanomaterial immunosafety by evaluating innate immunity across living species. *Small, 16*(21), 2000598. https://doi.org/10.1002/smll.202000598

Breitenbach Barroso Coelho, L. C., Marcelino dos Santos Silva, P., Felix de Oliveira, W., de Moura, M. C., Viana Pontual, E., Soares Gomes, F., Guedes Paiva, P. M., Napoleão, T. H., & dos Santos Correia, M. T. (2018). Lectins as antimicrobial agents. *Journal of Applied Microbiology, 125*(5), 1238–1252. https://doi.org/10.1111/jam.14055

Bricknell, I., & Dalmo, R. A. (2005). The use of immunostimulants in fish larval aquaculture. *Fish & Shellfish Immunology, 19*(5), 457–472. https://doi.org/10.1016/j.fsi.2005.03.008

Buchmann, K. (2014). Evolution of innate immunity: Clues from invertebrates via fish to mammals. *Frontiers in Immunology, 5*, 459.

Buchmann, K. (2018). Evolution of immunity. In E. L. Cooper (Ed.), *Advances in comparative immunology* (pp. 3–22). Springer International Publishing. https://doi.org/10.1007/978-3-319-76768-0_1

Caamaño, J., & Hunter, C. A. (2002). NF-κB family of transcription factors: Central regulators of innate and adaptive immune functions. *Clinical Microbiology Reviews, 15*(3), 414–429. https://doi.org/10.1128/cmr.15.3.414-429.2002

Cammarata-Mouchtouris, A., Acker, A., Goto, A., Chen, D., Matt, N., & Leclerc, V. (2022). Dynamic regulation of NF-κB response in innate immunity: The case of the IMD pathway in drosophila. *Biomedicine, 10*(9), Article 9. https://doi.org/10.3390/biomedicines10092304

Cappellano, G., Carecchio, M., Fleetwood, T., Magistrelli, L., Cantello, R., Dianzani, U., & Comi, C. (2013). Immunity and inflammation in neurodegenerative diseases. *American Journal of Neurodegenerative Disease, 2*(2), 89–107.

Cerenius, L., & Söderhäll, K. (2021). Immune properties of invertebrate phenoloxidases. *Developmental & Comparative Immunology, 122*, 104098. https://doi.org/10.1016/j.dci.2021.104098

Cerenius, L., Lee, B. L., & Söderhäll, K. (2008). The proPO-system: Pros and cons for its role in invertebrate immunity. *Trends in Immunology, 29*(6), 263–271. https://doi.org/10.1016/j.it.2008.02.009

Černý, J., & Stříž, I. (2019). Adaptive innate immunity or innate adaptive immunity? *Clinical Science, 133*(14), 1549–1565. https://doi.org/10.1042/CS20180548

Chainy, G. B. N., Paital, B., & Dandapat, J. (2016). An overview of seasonal changes in oxidative stress and antioxidant defence parameters in some invertebrate and vertebrate species. *Scientifica, 2016*(1), 6126570. https://doi.org/10.1155/2016/6126570

Chikezie, F. M., Opara, K. N., & Ubulom, P. M. E. (2024). Impacts of changing climate on arthropod vectors and diseases transmission. *Nigerian Journal of Entomology, 40*, 179–192.

Ciancio, A. (2016). Defense and immune systems. In A. Ciancio (Ed.), *Invertebrate bacteriology: Function, evolution and biological ties* (pp. 205–239). Springer. https://doi.org/10.1007/978-94-024-0884-3_7

Clark, K. F., & Greenwood, S. J. (2016). Next-generation sequencing and the crustacean immune system: The need for alternatives in immune gene annotation. *Integrative and Comparative Biology, 56*(6), 1113–1130. https://doi.org/10.1093/icb/icw023

Clark, M. S., Peck, L. S., Arivalagan, J., Backeljau, T., Berland, S., Cardoso, J. C. R., Caurcel, C., Chapelle, G., De Noia, M., Dupont, S., Gharbi, K., Hoffman, J. I., Last, K. S., Marie, A., Melzner, F., Michalek, K., Morris, J., Power, D. M., Ramesh, K., et al. (2020). Deciphering mollusc shell production: The roles of genetic mechanisms through to ecology, aquaculture and biomimetics. *Biological Reviews, 95*(6), 1812–1837. https://doi.org/10.1111/brv.12640

Coates, C. J., Rowley, A. F., Smith, L. C., & Whitten, M. M. (2022). Host defences of invertebrates to pathogens and parasites. *Invertebrate Pathology, 1.* https://www.researchgate.net/profile/L-Smith-8/publication/358404400_Host_defences_of_invertebrates_to_pathogens_and_parasites/links/62129e79eb735c508ae50fed/Host-defences-of-invertebrates-to-pathogens-and-parasites.pdf

Cooper, E. L. (2010). Evolution of immune systems from self/not self to danger to artificial immune systems (AIS). *Physics of Life Reviews, 7*(1), 55–78. https://doi.org/10.1016/j.plrev.2009.12.001

Cooper, D., & Eleftherianos, I. (2017). Memory and specificity in the insect immune system: Current perspectives and future challenges. *Frontiers in Immunology, 8.* https://doi.org/10.3389/fimmu.2017.00539

Destoumieux-Garzón, D., Rosa, R. D., Schmitt, P., Barreto, C., Vidal-Dupiol, J., Mitta, G., Gueguen, Y., & Bachère, E. (2016). Antimicrobial peptides in marine invertebrate health and disease. *Philosophical Transactions of the Royal Society B: Biological Sciences, 371*(1695), 20150300. https://doi.org/10.1098/rstb.2015.0300

Dhinaut, J., Chogne, M., & Moret, Y. (2018). Immune priming specificity within and across generations reveals the range of pathogens affecting evolution of immunity in an insect. *Journal of Animal Ecology, 87*(2), 448–463. https://doi.org/10.1111/1365-2656.12661

Dong, C., Wei, Z., & Yang, G. (2011). Involvement of peroxinectin in the defence of red swamp crayfish Procambarus clarkii against pathogenic Aeromonas hydrophila. *Fish & Shellfish Immunology, 30*(6), 1223–1229.

Du Pasquier, L. (2001). The immune system of invertebrates and vertebrates. *Comparative Biochemistry and Physiology Part B: Biochemistry and Molecular Biology, 129*(1), 1–15. https://doi.org/10.1016/S1096-4959(01)00306-2

Dubief, B., Nunes, F. L. D., Basuyaux, O., & Paillard, C. (2017). Immune priming and portal of entry effectors improve response to vibrio infection in a resistant population of the European abalone. *Fish & Shellfish Immunology, 60*, 255–264. https://doi.org/10.1016/j.fsi.2016.11.017

Eleftherianos, I., Heryanto, C., Bassal, T., Zhang, W., Tettamanti, G., & Mohamed, A. (2021). Haemocyte-mediated immunity in insects: Cells, processes and associated components in the fight against pathogens and parasites. *Immunology, 164*(3), 401–432. https://doi.org/10.1111/imm.13390

Endo, Y., Matsushita, M., & Fujita, T. (2015). Chapter Two—New insights into the role of Ficolins in the lectin pathway of innate immunity. In K. W. Jeon (Ed.), *International review of cell and molecular biology* (Vol. 316, pp. 49–110). Academic. https://doi.org/10.1016/bs.ircmb.2015.01.003

Engelmann, P., Cooper, E. L., Opper, B., & Németh, P. (2011). Earthworm innate immune system. In A. Karaca (Ed.), *Biology of earthworms* (pp. 229–245). Springer. https://doi.org/10.1007/978-3-642-14636-7_14

Ferraboschi, P., Ciceri, S., & Grisenti, P. (2021). Applications of lysozyme, an innate immune Defense factor, as an alternative antibiotic. *Antibiotics, 10*(12), Article 12. https://doi.org/10.3390/antibiotics10121534

Frøystad, M. K., Volden, V., Berg, T., & Gjøen, T. (2002). Metabolism of oxidized and chemically modified low density lipoproteins in rainbow trout—Clearance via scavenger receptors. *Developmental & Comparative Immunology, 26*(8), 723–733.

Fujita, T. (2002). Evolution of the lectin–complement pathway and its role in innate immunity. *Nature Reviews Immunology, 2*(5), 346–353. https://doi.org/10.1038/nri800

Gatto, M., Iaccarino, L., Ghirardello, A., Bassi, N., Pontisso, P., Punzi, L., Shoenfeld, Y., & Doria, A. (2013). Serpins, immunity and autoimmunity: Old molecules, new functions. *Clinical Reviews in Allergy & Immunology, 45*(2), 267–280. https://doi.org/10.1007/s12016-013-8353-3

Gourbal, B., Pinaud, S., Beckers, G. J. M., Van Der Meer, J. W. M., Conrath, U., & Netea, M. G. (2018). Innate immune memory: An evolutionary perspective. *Immunological Reviews, 283*(1), 21–40. https://doi.org/10.1111/imr.12647

Greenlee-Wacker, M. C. (2016). Clearance of apoptotic neutrophils and resolution of inflammation. *Immunological Reviews, 273*(1), 357–370. https://doi.org/10.1111/imr.12453

Guo, X., He, Y., Zhang, L., Lelong, C., & Jouaux, A. (2015). Immune and stress responses in oysters with insights on adaptation. *Fish & Shellfish Immunology, 46*(1), 107–119. https://doi.org/10.1016/j.fsi.2015.05.018

Gupta, A. P. (1991). Insect Immunocytes and other Hemocytes: Roles in cellular and humoral immunity*. In *Immunology of insects and other arthropods*. CRC Press.

Guryanova, S. V., Balandin, S. V., Belogurova-Ovchinnikova, O. Y., & Ovchinnikova, T. V. (2023). Marine invertebrate antimicrobial peptides and their potential as novel peptide antibiotics. *Marine Drugs, 21*(10), Article 10. https://doi.org/10.3390/md21100503

Habib, Y. J., & Zhang, Z. (2020). The involvement of crustaceans toll-like receptors in pathogen recognition. *Fish & Shellfish Immunology, 102*, 169–176. https://doi.org/10.1016/j.fsi.2020.04.035

Hanson, M. A., & Lemaitre, B. (2020). New insights on *Drosophila* antimicrobial peptide function in host defense and beyond. *Current Opinion in Immunology, 62*, 22–30. https://doi.org/10.1016/j.coi.2019.11.008

Huang, Y., & Ren, Q. (2020). Research progress in innate immunity of freshwater crustaceans. *Developmental & Comparative Immunology, 104*, 103569. https://doi.org/10.1016/j.dci.2019.103569

Iatsenko, I., Kondo, S., Mengin-Lecreulx, D., & Lemaitre, B. (2016). PGRP-SD, an extracellular pattern-recognition receptor, enhances peptidoglycan-mediated activation of the Drosophila Imd pathway. *Immunity, 45*(5), 1013–1023.

Imler, J.-L. (2014). Overview of *Drosophila* immunity: A historical perspective. *Developmental & Comparative Immunology, 42*(1), 3–15. https://doi.org/10.1016/j.dci.2013.08.018

Ito, S., Sugumaran, M., & Wakamatsu, K. (2020). Chemical reactivities of ortho-quinones produced in living organisms: Fate of quinonoid products formed by tyrosinase and phenoloxidase action on phenols and catechols. *International Journal of Molecular Sciences, 21*(17), Article 17. https://doi.org/10.3390/ijms21176080

Iwasaki, A., & Medzhitov, R. (2015). Control of adaptive immunity by the innate immune system. *Nature Immunology, 16*(4), 343–353. https://doi.org/10.1038/ni.3123

Jack, R., & Du Pasquier, L. (2019). Innate immunity. In R. Jack & L. Du Pasquier (Eds.), *Evolutionary concepts in immunology* (pp. 33–69). Springer International Publishing. https://doi.org/10.1007/978-3-030-18667-8_3

Janik-Karpinska, E., Ceremuga, M., Niemcewicz, M., Podogrocki, M., Stela, M., Cichon, N., & Bijak, M. (2022). Immunosensors—The future of pathogen real-time detection. *Sensors, 22*(24), Article 24. https://doi.org/10.3390/s22249757

Janssens, S., & Beyaert, R. (2003). Role of toll-like receptors in pathogen recognition. *Clinical Microbiology Reviews, 16*(4), 637–646. https://doi.org/10.1128/cmr.16.4.637-646.2003

Jearaphunt, M., Noonin, C., Jiravanichpaisal, P., Nakamura, S., Tassanakajon, A., Söderhäll, I., & Söderhäll, K. (2014). Caspase-1-like regulation of the proPO-system and role of ppA and Caspase-1-like cleaved peptides from proPO in innate immunity. *PLoS Pathogens, 10*(4), e1004059. https://doi.org/10.1371/journal.ppat.1004059

Jia, Z., Zhang, H., Jiang, S., Wang, M., Wang, L., & Song, L. (2016). Comparative study of two single CRD C-type lectins, CgCLec-4 and CgCLec-5, from pacific oyster Crassostrea gigas. *Fish & Shellfish Immunology, 59*, 220–232.

Katzenback, B. A. (2015). Antimicrobial peptides as mediators of innate immunity in Teleosts. *Biology, 4*(4), Article 4. https://doi.org/10.3390/biology4040607

Kaufmann, S. H. E., Rouse, B. T., & Sacks, D. L. (2010). *The immune response to infection.* American Society for Microbiology Press.

Kleino, A., Myllymäki, H., Kallio, J., Vanha-aho, L.-M., Oksanen, K., Ulvila, J., Hultmark, D., Valanne, S., & Rämet, M. (2008). Pirk is a negative regulator of the Drosophila Imd Pathway1. *The Journal of Immunology, 180*(8), 5413–5422. https://doi.org/10.4049/jimmunol.180.8.5413

Kulkarni, A., Krishnan, S., Anand, D., Kokkattunivarthil Uthaman, S., Otta, S. K., Karunasagar, I., & Kooloth Valappil, R. (2021). Immune responses and immunoprotection in crustaceans with special reference to shrimp. *Reviews in Aquaculture, 13*(1), 431–459. https://doi.org/10.1111/raq.12482

Kumar, S., Verma, A. K., Singh, S. P., & Awasthi, A. (2023). Immunostimulants for shrimp aquaculture: Paving pathway towards shrimp sustainability. *Environmental Science and Pollution Research, 30*(10), 25325–25343. https://doi.org/10.1007/s11356-021-18433-y

Kumar, S., Mondal, K., Thakur, N., & Das, S. (2024). 1 polyphenol oxidases: An enzyme of bacteria and fungi. In P. Verma, K. Agrawal, & M. P. Shah (Eds.), *Polyphenol oxidases: Function, wastewater remediation, and biosensors* (pp. 1–24). De Gruyter. https://www.degruyterbrill.com/document/doi/10.1515/9783111033525-001/pdf?licenseType=restricted

Lavine, M. D., & Strand, M. R. (2002). Insect hemocytes and their role in immunity. *Insect Biochemistry and Molecular Biology, 32*(10), 1295–1309. https://doi.org/10.1016/S0965-1748(02)00092-9

Lazzaro, B. P., Zasloff, M., & Rolff, J. (2020). Antimicrobial peptides: Application informed by evolution. *Science, 368*(6490), eaau5480. https://doi.org/10.1126/science.aau5480

Lemaitre, B., & Hoffmann, J. (2007a). The host defense of Drosophila melanogaster. *Annual Review of Immunology, 25*(1), 697–743.

Li, P., & Chang, M. (2021). Roles of PRR-mediated signaling pathways in the regulation of oxidative stress and inflammatory diseases. *International Journal of Molecular Sciences, 22*(14), Article 14. https://doi.org/10.3390/ijms22147688

Li, D., & Wu, M. (2021). Pattern recognition receptors in health and diseases. *Signal Transduction and Targeted Therapy, 6*(1), 1–24. https://doi.org/10.1038/s41392-021-00687-0

Li, F., & Xiang, J. (2013). Signaling pathways regulating innate immune responses in shrimp. *Fish & Shellfish Immunology, 34*(4), 973–980. https://doi.org/10.1016/j.fsi.2012.08.023

Li, H., Zhao, Q., Xu, J., Li, X., Chen, X., Zhang, Y., Li, H., Zhu, Y., Liu, M., Zhao, L., Hua, D., Zhang, X., & Chen, K. (2025). From Biomphalaria glabrata to Drosophila melanogaster and Anopheles gambiae: The diversity and role of FREPs and Dscams in immune response. *Frontiers in Immunology, 16*, 1579905. https://doi.org/10.3389/fimmu.2025.1579905

Little, T. J., & Kraaijeveld, A. R. (2004). Ecological and evolutionary implications of immunological priming in invertebrates. *Trends in Ecology & Evolution, 19*(2), 58–60. https://doi.org/10.1016/j.tree.2003.11.011

Liu, F.-T., & Stowell, S. R. (2023). The role of galectins in immunity and infection. *Nature Reviews Immunology, 23*(8), 479–494. https://doi.org/10.1038/s41577-022-00829-7

Liu, H., Söderhäll, K., & Jiravanichpaisal, P. (2009). Antiviral immunity in crustaceans. *Fish & Shellfish Immunology, 27*(2), 79–88. https://doi.org/10.1016/j.fsi.2009.02.009

Liu, S., Zheng, S.-C., Li, Y.-L., Li, J., & Liu, H.-P. (2020). Hemocyte-mediated phagocytosis in crustaceans. *Frontiers in Immunology, 11.* https://doi.org/10.3389/fimmu.2020.00268

Loh, B., Gondil, V. S., Manohar, P., Khan, F. M., Yang, H., & Leptihn, S. (2021). Encapsulation and delivery of therapeutic phages. *Applied and Environmental Microbiology, 87*(5), e01979–e01920. https://doi.org/10.1128/AEM.01979-20

Loker, E. S., & Bayne, C. J. (2018). Molluscan immunobiology: Challenges in the anthropocene epoch. In E. L. Cooper (Ed.), *Advances in comparative immunology* (pp. 343–407). Springer International Publishing. https://doi.org/10.1007/978-3-319-76768-0_12

Macho, A. P., & Zipfel, C. (2014). Plant PRRs and the activation of innate immune Signaling. *Molecular Cell, 54*(2), 263–272. https://doi.org/10.1016/j.molcel.2014.03.028

Mahanta, D. K., Bhoi, T. K., Komal, J., Samal, I., Nikhil, R. M., Paschapur, A. U., Singh, G., Kumar, P. V. D., Desai, H. R., Ahmad, M. A., Singh, P. P., Majhi, P. K., Mukherjee, U., Singh, P., Saini, V., Shahanaz, S., & N., & Yele, Y. (2023). Insect-pathogen crosstalk and the cellular-molecular mechanisms of insect immunity: Uncovering the underlying signaling pathways and immune regulatory function of non-coding RNAs. *Frontiers in Immunology, 14.* https://doi.org/10.3389/fimmu.2023.1169152

Mba, I. E., & Nweze, E. I. (2022). Antimicrobial peptides therapy: An emerging alternative for treating drug-resistant bacteria. *The Yale Journal of Biology and Medicine, 95*(4), 445–463.

McNeela, E. A., & Mills, K. H. G. (2001). Manipulating the immune system: Humoral versus cell-mediated immunity. *Advanced Drug Delivery Reviews, 51*(1), 43–54. https://doi.org/10.1016/S0169-409X(01)00169-7

Melillo, D., Marino, R., Italiani, P., & Boraschi, D. (2018). Innate immune memory in invertebrate metazoans: A critical appraisal. *Frontiers in Immunology, 9.* https://doi.org/10.3389/fimmu.2018.01915

Montagnani, C., Morga, B., Novoa, B., Gourbal, B., Saco, A., Rey-Campos, M., Bourhis, M., Riera, F., Vignal, E., Corporeau, C., Charrière, G. M., Travers, M.-A., Dégremont, L., Gueguen, Y., Cosseau, C., & Figueras, A. (2024). Trained immunity: Perspectives for disease control strategy in marine mollusc aquaculture. *Reviews in Aquaculture, 16*(4), 1472–1498. https://doi.org/10.1111/raq.12906

Müller, L., Fülöp, T., & Pawelec, G. (2013). Immunosenescence in vertebrates and invertebrates. *Immunity & Ageing, 10*(1), 12. https://doi.org/10.1186/1742-4933-10-12

Myllymäki, H., & Rämet, M. (2014). JAK/STAT pathway in drosophila immunity. *Scandinavian Journal of Immunology, 79*(6), 377–385. https://doi.org/10.1111/sji.12170

Nabi-Afjadi, M., Heydari, M., Zalpoor, H., Arman, I., Sadoughi, A., Sahami, P., & Aghazadeh, S. (2022). Lectins and lectibodies: Potential promising antiviral agents. *Cellular & Molecular Biology Letters, 27*, 37. https://doi.org/10.1186/s11658-022-00338-4

Nappi, A. J., & Ottaviani, E. (2000). Cytotoxicity and cytotoxic molecules in invertebrates. *BioEssays, 22*(5), 469–480. https://doi.org/10.1002/(SICI)1521-1878(200005)22:5<469::AID-BIES9>3.0.CO;2-4

Ng, T. H., & Kurtz, J. (2020). Dscam in immunity: A question of diversity in insects and crustaceans. *Developmental & Comparative Immunology, 105*, 103539. https://doi.org/10.1016/j.dci.2019.103539

Ng, T. H., Chiang, Y.-A., Yeh, Y.-C., & Wang, H.-C. (2014). Review of Dscam-mediated immunity in shrimp and other arthropods. *Developmental & Comparative Immunology, 46*(2), 129–138. https://doi.org/10.1016/j.dci.2014.04.002

Nguyen, N. H. (2024). Genetics and genomics of infectious diseases in key aquaculture species. *Biology, 13*(1), Article 1. https://doi.org/10.3390/biology13010029

Nie, L., Cai, S.-Y., Shao, J.-Z., & Chen, J. (2018). Toll-like receptors, associated biological roles, and signaling networks in non-mammals. *Frontiers in Immunology, 9.* https://doi.org/10.3389/fimmu.2018.01523

Nizamani, M. M., Zhang, Q., Muhae-Ud-Din, G., & Wang, Y. (2023). High-throughput sequencing in plant disease management: A comprehensive review of benefits, challenges, and future perspectives. *Phytopathology Research, 5*(1), 44. https://doi.org/10.1186/s42483-023-00199-5

Okoro, O. J., Deme, G. G., Okoye, C. O., Eze, S. C., Odii, E. C., Gbadegesin, J. T., Okeke, E. S., Oyejobi, G. K., Nyaruaba, R., & Ebido, C. C. (2023). Understanding key vectors and vector-borne diseases associated with freshwater ecosystem across Africa: Implications for public health. *Science of the Total Environment, 862*, 160732. https://doi.org/10.1016/j.scitotenv.2022.160732

Orlov, D. S., Shamova, O. V., Eliseev, I. E., Zharkova, M. S., Chakchir, O. B., Antcheva, N., et al. (2019). Redesigning arenicin-1, an antimicrobial peptide from the marine polychaeta Arenicola marina, by strand rearrangement or branching, substitution of specific residues, and backbone linearization or cyclization. *Marine Drugs, 17*(6), 376.

Parisi, M. G., Parrinello, D., Stabili, L., & Cammarata, M. (2020). Cnidarian immunity and the repertoire of defense mechanisms in anthozoans. *Biology, 9*(9), 283.

Patnaik, B. B., Baliarsingh, S., Sarkar, A., Hameed, A. S. S., Lee, Y. S., Jo, Y. H., Han, Y. S., & Mohanty, J. (2024). The role of pattern recognition receptors in crustacean innate immunity. *Reviews in Aquaculture, 16*(1), 190–233. https://doi.org/10.1111/raq.12829

Pees, B., Yang, W., Zárate-Potes, A., Schulenburg, H., & Dierking, K. (2015). High innate immune specificity through diversified C-type lectin-like domain proteins in invertebrates. *Journal of Innate Immunity, 8*(2), 129–142. https://doi.org/10.1159/000441475

Perveen, N., Muhammad, K., Muzaffar, S. B., Zaheer, T., Munawar, N., Gajic, B., Sparagano, O. A., Kishore, U., & Willingham, A. L. (2023). Host-pathogen interaction in arthropod vectors: Lessons from viral infections. *Frontiers in Immunology, 14*. https://doi.org/10.3389/fimmu.2023.1061899

Pfalzgraff, A., Brandenburg, K., & Weindl, G. (2018). Antimicrobial peptides and their therapeutic potential for bacterial skin infections and wounds. *Frontiers in Pharmacology, 9*, 281. https://doi.org/10.3389/fphar.2018.00281

Prakash, A., Fenner, F., Shit, B., Salminen, T. S., Monteith, K. M., Khan, I., & Vale, P. F. (2024). IMD-mediated innate immune priming increases Drosophila survival and reduces pathogen transmission. *PLoS Pathogens, 20*(6), e1012308. https://doi.org/10.1371/journal.ppat.1012308

Pukkila-Worley, R., Feinbaum, R., Kirienko, N. V., Larkins-Ford, J., Conery, A. L., & Ausubel, F. M. (2012). Stimulation of host immune defenses by a small molecule protects C. elegans from bacterial infection. *PLoS Genetics, 8*(6), e1002733.

Pull, C. D., & McMahon, D. P. (2020). Superorganism immunity: A major transition in immune system evolution. *Frontiers in Ecology and Evolution, 8*. https://doi.org/10.3389/fevo.2020.00186

Radtke, F., Fasnacht, N., & MacDonald, H. R. (2010). Notch signaling in the immune system. *Immunity, 32*(1), 14–27. https://doi.org/10.1016/j.immuni.2010.01.004

Ragland, S. A., & Criss, A. K. (2017). From bacterial killing to immune modulation: Recent insights into the functions of lysozyme. *PLoS Pathogens, 13*(9), e1006512. https://doi.org/10.1371/journal.ppat.1006512

Rathinam, R. B., Acharya, A., Robina, A. J., Banu, H., & Tripathi, G. (2024). The immune system of marine invertebrates: Earliest adaptation of animals. *Comparative Immunology Reports, 7*, 200163. https://doi.org/10.1016/j.cirep.2024.200163

Rauta, P. R., Samanta, M., Dash, H. R., Nayak, B., & Das, S. (2014). Toll-like receptors (TLRs) in aquatic animals: Signaling pathways, expressions and immune responses. *Immunology Letters, 158*(1), 14–24. https://doi.org/10.1016/j.imlet.2013.11.013

Riera Romo, M., Pérez-Martínez, D., & Castillo Ferrer, C. (2016). Innate immunity in vertebrates: An overview. *Immunology, 148*(2), 125–139. https://doi.org/10.1111/imm.12597

Rodrigues, T., Guardiola, F. A., Almeida, D., & Antunes, A. (2025). Aquatic invertebrate antimicrobial peptides in the fight against aquaculture pathogens. *Microorganisms, 13*(1), Article 1. https://doi.org/10.3390/microorganisms13010156

Romero, A., Dios, S., Poisa-Beiro, L., Costa, M. M., Posada, D., Figueras, A., & Novoa, B. (2011). Individual sequence variability and functional activities of fibrinogen-related proteins (FREPs) in the Mediterranean mussel (Mytilus galloprovincialis) suggest ancient and complex immune recognition models in invertebrates. *Developmental & Comparative Immunology, 35*(3), 334–344.

Rosa, R. D., & Barracco, M. A. (2010). Antimicrobial peptides in crustaceans. *Invertebrate Survival Journal, 7*(2), Article 2.

Rowley, A. F., Coates, C. J., & Whitten, M. M. (2022). *Invertebrate pathology*. Oxford University Press. https://books.google.com/books?hl=en&lr=&id=_6FaEAAAQBAJ&oi=fnd&pg=PP1&dq=Encapsulation+is+a+highly+coordinated+and+cooperative+immune+response+widely+observed+in+insects+and+some+crustaceans,+serving+as+an+essential+mechanism+against+macroparasites+and+other+large+foreign+bodies&ots=-pVZ5vGQm-&sig=QMG6vtsCNSMBdpn5aioiHi1XdAo.

Roy, S., Bossier, P., Norouzitallab, P., & Vanrompay, D. (2020). Trained immunity and perspectives for shrimp aquaculture. *Reviews in Aquaculture, 12*(4), 2351–2370. https://doi.org/10.1111/raq.12438

Sacchi, S., Malagoli, D., & Franchi, N. (2024). The invertebrate Immunocyte: A complex and versatile model for immunological, developmental, and environmental research. *Cells, 13*(24), Article 24. https://doi.org/10.3390/cells13242106

Salam, M. A., Al-Amin, M. Y., Salam, M. T., Pawar, J. S., Akhter, N., Rabaan, A. A., & Alqumber, M. A. A. (2023). Antimicrobial resistance: A growing serious threat for global public health. *Healthcare, 11*(13), Article 13. https://doi.org/10.3390/healthcare11131946

Salcedo-Porras, N., Noor, S., Cai, C., Oliveira, P. L., & Lowenberger, C. (2021). *Rhodnius prolixus* uses the peptidoglycan recognition receptor rpPGRP-LC/LA to detect gram-negative bacteria and activate the IMD pathway. *Current Research in Insect Science, 1*, 100006. https://doi.org/10.1016/j.cris.2020.100006

Salvador, A. F., de Lima, K. A., & Kipnis, J. (2021). Neuromodulation by the immune system: A focus on cytokines. *Nature Reviews Immunology, 21*(8), 526–541. https://doi.org/10.1038/s41577-021-00508-z

Satyavathi, V. V., Minz, A., & Nagaraju, J. (2014). Nodulation: An unexplored cellular defense mechanism in insects. *Cellular Signalling, 26*(8), 1753–1763. https://doi.org/10.1016/j.cellsig.2014.02.024

Schaefer, A. K., Melnyk, J. E., He, Z., Del Rosario, F., & Grimes, C. L. (2018). Chapter 14—Pathogen- and microbial- associated molecular patterns (PAMPs/MAMPs) and the innate immune response in Crohn's disease. In S. Chatterjee, W. Jungraithmayr, & D. Bagchi (Eds.), *Immunity and inflammation in health and disease* (pp. 175–187). Academic. https://doi.org/10.1016/B978-0-12-805417-8.00014-7

Schneider, J. (2022). *Characterization of the antiviral STING pathway in "Drosophila melanogaster": Signalling and NF-κB factor activation* [Phdthesis, Université de Strasbourg]. https://theses.hal.science/tel-04416425

Schulenburg, H., Léopold Kurz, C., & Ewbank, J. J. (2004). Evolution of the innate immune system: The worm perspective. *Immunological Reviews, 198*(1), 36–58. https://doi.org/10.1111/j.0105-2896.2004.0125.x

Sekino, M., Sato, S. I., Hong, J. S., & Li, Q. (2012). Contrasting pattern of mitochondrial population diversity between an estuarine bivalve, the Kumamoto oyster Crassostrea sikamea, and the closely related Pacific oyster C. gigas. *Marine Biology, 159*(12), 2757–2776.

Sharrock, J., & Sun, J. C. (2020). Innate immunological memory: From plants to animals. *Current Opinion in Immunology, 62*, 69–78. https://doi.org/10.1016/j.coi.2019.12.001

Sheehan, G., Farrell, G., & Kavanagh, K. (2020). Immune priming: The secret weapon of the insect world. *Virulence, 11*(1), 238–246. https://doi.org/10.1080/21505594.2020.1731137

Silva, R. C. M. C., & Gomes, F. M. (2024). Evolution of the major components of innate immunity in animals. *Journal of Molecular Evolution, 92*(1), 3–20. https://doi.org/10.1007/s00239-024-10155-2

Silva-Gomes, S., Decout, A., & Nigou, J. (2014). Pathogen-associated molecular patterns (PAMPs). In *Encyclopedia of inflammatory diseases* (pp. 1–16). Birkhäuser. https://doi.org/10.1007/978-3-0348-0620-6_35-1

Sindelar, R. D. (2024). Genomics, other "OMIC" technologies, precision medicine, and additional biotechnology-related techniques. In D. J. A. Crommelin, R. D. Sindelar, & B. Meibohm (Eds.), *Pharmaceutical biotechnology: Fundamentals and applications* (pp. 209–254). Springer International Publishing. https://doi.org/10.1007/978-3-031-30023-3_9

Soares, M. P., Teixeira, L., & Moita, L. F. (2017). Disease tolerance and immunity in host protection against infection. *Nature Reviews Immunology, 17*(2), 83–96. https://doi.org/10.1038/nri.2016.136

Söderhäll, K. (2024). Invertebrate immunology—Some thoughts about past and future research. *Developmental & Comparative Immunology, 161*, 105256. https://doi.org/10.1016/j.dci.2024.105256

Steiner, H. (2004). Peptidoglycan recognition proteins: On and off switches for innate immunity. *Immunological Reviews, 198*(1), 83–96. https://doi.org/10.1111/j.0105-2896.2004.0120.x

Sułek, M., Kordaczuk, J., & Wojda, I. (2021). Current understanding of immune priming phenomena in insects. *Journal of Invertebrate Pathology, 185*, 107656. https://doi.org/10.1016/j.jip.2021.107656

Tassanakajon, A., Somboonwiwat, K., & Amparyup, P. (2015). Sequence diversity and evolution of antimicrobial peptides in invertebrates. *Developmental & Comparative Immunology, 48*(2), 324–341. https://doi.org/10.1016/j.dci.2014.05.020

Tosi, M. F. (2005). Innate immune responses to infection. *Journal of Allergy and Clinical Immunology, 116*(2), 241–249. https://doi.org/10.1016/j.jaci.2005.05.036

Tran, N. T., Liang, H., Zhang, M., Bakky, M. A. H., Zhang, Y., & Li, S. (2022). Role of cellular receptors in the innate immune system of crustaceans in response to White Spot syndrome virus. *Viruses, 14*(4), 743. https://doi.org/10.3390/v14040743

Uengwetwanit, T., Uawisetwathana, U., Angthong, P., Phanthura, M., Phromson, M., Tala, S., Thepsuwan, T., Chaiyapechara, S., Prathumpai, W., & Rungrassamee, W. (2025). Investigating a novel β-glucan source to enhance disease resistance in Pacific white shrimp (Penaeus vannamei). *Scientific Reports, 15*, 15377. https://doi.org/10.1038/s41598-025-00157-5

Vallabhapurapu, S., & Karin, M. (2009). Regulation and function of NF-κB transcription factors in the immune system. *Annual Review of Immunology, 27*, 693–733. https://doi.org/10.1146/annurev.immunol.021908.132641

Vasta, G. R., Ahmed, H., Tasumi, S., Odom, E. W., & Saito, K. (2007). Biological roles of lectins in innate immunity: Molecular and structural basis for diversity in self/non-self recognition. In J. D. Lambris (Ed.), *Current topics in innate immunity* (pp. 389–406). Springer. https://doi.org/10.1007/978-0-387-71767-8_27

Voogdt, C. G. P., & van Putten, J. P. M. (2016). Chapter 13—The evolution of the Toll-like receptor system. In D. Malagoli (Ed.), *The evolution of the immune system* (pp. 311–330). Academic. https://doi.org/10.1016/B978-0-12-801975-7.00013-X

Wang, X.-W., & Wang, J.-X. (2013). Pattern recognition receptors acting in innate immune system of shrimp against pathogen infections. *Fish & Shellfish Immunology, 34*(4), 981–989. https://doi.org/10.1016/j.fsi.2012.08.008

Wang, L., Zhang, H., Wang, M., Zhou, Z., Wang, W., Liu, R., Huang, M., Yang, C., Qiu, L., & Song, L. (2019a). The transcriptomic expression of pattern recognition receptors: Insight into molecular recognition of various invading pathogens in Oyster *Crassostrea gigas*. *Developmental & Comparative Immunology, 91*, 1–7. https://doi.org/10.1016/j.dci.2018.09.021

Wang, X., Zhang, Y., Zhang, R., & Zhang, J. (2019b). The diversity of pattern recognition receptors (PRRs) involved with insect defense against pathogens. *Current Opinion in Insect Science, 33*, 105–110. https://doi.org/10.1016/j.cois.2019.05.004

Wang, X. W., Vasta, G. R., & Wang, J. X. (2020). The functional relevance of shrimp C-type lectins in host-pathogen interactions. *Developmental & Comparative Immunology, 109*, 103708.

Wang, W., Wang, L., & Song, L. (2025). The immune priming in aquaculture invertebrates: Inspiration from cellular perspective and future investigation. *Reviews in Aquaculture, 17*(1), e12977. https://doi.org/10.1111/raq.12977

Wicherska-Pawłowska, K., Wróbel, T., & Rybka, J. (2021). Toll-like receptors (TLRs), NOD-like receptors (NLRs), and RIG-I-like receptors (RLRs) in innate immunity. TLRs, NLRs, and RLRs ligands as immunotherapeutic agents for hematopoietic diseases. *International Journal of Molecular Sciences, 22*(24), Article 24. https://doi.org/10.3390/ijms222413397

Woods, N., Niwasabutra, K., Acevedo, R., Igoli, J., Altwaijry, N. A., Tusiimire, J., Gray, A. I., Watson, D. G., & Ferro, V. A. (2017). Chapter 11—Natural vaccine adjuvants and immunopotentiators derived from plants, fungi, marine organisms, and insects. In V. E. J. C. Schijns & D. T. O'Hagan (Eds.), *Immunopotentiators in modern vaccines* (2nd ed., pp. 211–229). Academic. https://doi.org/10.1016/B978-0-12-804019-5.00011-6

Yamamoto, M., & Takeda, K. (2008). Role of nuclear IκB proteins in the regulation of host immune responses. *Journal of Infection and Chemotherapy, 14*(4), 265–269. https://doi.org/10.1007/s10156-008-0619-y

Young, N. D., Jex, A. R., Li, B., Liu, S., Yang, L., Xiong, Z., et al. (2012). Whole-genome sequence of Schistosoma haematobium. *Nature Genetics, 44*(2), 221–225.

Zambon, R. A., Nandakumar, M., Vakharia, V. N., & Wu, L. P. (2005). The Toll pathway is important for an antiviral response in Drosophila. *Proceedings of the National Academy of Sciences, 102*(20), 7257–7262. https://doi.org/10.1073/pnas.0409181102

Zhai, Z., Boquete, J.-P., & Lemaitre, B. (2018). Cell-specific Imd-NF-κB responses enable simultaneous antibacterial immunity and intestinal epithelial cell shedding upon bacterial infection. *Immunity, 48*(5), 897–910.e7. https://doi.org/10.1016/j.immuni.2018.04.010

Zhang, G., & Ghosh, S. (2001). Toll-like receptor–mediated NF-κB activation: A phylogenetically conserved paradigm in innate immunity. *The Journal of Clinical Investigation, 107*(1), 13–19. https://doi.org/10.1172/JCI11837

Zhang, L., Li, L., & Zhang, G. (2011). A *Crassostrea gigas* Toll-like receptor and comparative analysis of TLR pathway in invertebrates. *Fish & Shellfish Immunology, 30*(2), 653–660. https://doi.org/10.1016/j.fsi.2010.12.023

Zhang, Z.-L., Meng, Y.-Q., Li, J.-J., Zhang, X.-X., Li, J.-T., Xu, J.-R., Zheng, P.-H., Xian, J.-A., & Lu, Y.-P. (2024). Effects of antimicrobial peptides from dietary *Hermetia illucens* larvae on the growth, immunity, gene expression, intestinal microbiota and resistance to *Aeromonas hydrophila* of juvenile red claw crayfish (*Cherax quadricarinatus*). *Fish & Shellfish Immunology, 147*, 109437. https://doi.org/10.1016/j.fsi.2024.109437

Zhao, B.-R., Wang, X.-X., Liu, P.-P., & Wang, X.-W. (2023). Complement-related proteins in crustacean immunity. *Developmental & Comparative Immunology, 139*, 104577. https://doi.org/10.1016/j.dci.2022.104577

Zhong, T.-Y., Arancibia, S., Born, R., Tampe, R., Villar, J., Del Campo, M., Manubens, A., & Becker, M. I. (2016). Hemocyanins stimulate innate immunity by inducing different temporal patterns of proinflammatory cytokine expression in macrophages. *The Journal of Immunology Author Choice, 196*(11), 4650–4662. https://doi.org/10.4049/jimmunol.1501156

Zhou, Y.-L., Wang, L.-Z., Gu, W.-B., Wang, C., Zhu, Q.-H., Liu, Z.-P., Chen, Y.-Y., & Shu, M.-A. (2018). Identification and functional analysis of *immune deficiency* (IMD) from *Scylla paramamosain*: The first evidence of IMD signaling pathway involved in immune defense against bacterial infection in crab species. *Fish & Shellfish Immunology, 81*, 150–160. https://doi.org/10.1016/j.fsi.2018.07.016

Chapter 2
Surveillance and Monitoring of Cellular-mediated Immune Responses

2.1 Introduction

Over 95% of animal species are invertebrates, whose immune systems are extremely diverse and have evolved to be very efficient in protecting the animals from pathogens. Unlike vertebrates that arm themselves with both innate and adaptive immune systems, invertebrates have to rely on innate immunity that, although well developed, is also well developed and sophisticated (Coates et al., 2022). Humoral and cellular responses of immune systems are broadly classified into broad ways of detecting and eliminating pathogens. Immune mechanisms in invertebrates are very important in the recognition of pathogens and the initiation of appropriate responses and mainly consist of the detection of pathogen-associated molecular patterns (PAMPs) (Patnaikl et al., 2024). Pattern recognition receptors (PRRs) such as these PAMPs will trigger immune signaling pathways and set in motion various defense mechanisms (Betancourt et al., 2024). Invertebrate immune response plays an important role in their survival in different environments and reveals a more comprehensive understanding of the evolution of immune systems among animals.

2.2 Hemocytes and Their Types: Granulocytes and Agranulocytes/Hyalinocytes in Invertebrates

Immune cells called hemocytes scan for pathogens and get rid of them by taking them in and surrounding them while producing ROS and AMPs. Immune system cells travel through the hemolymph fluid to guard the body across the insect's system (Eleftherianos et al., 2021). Hemocytes develop into two basic cell types by their physical traits and working methods. Granulocytes contain visible cell granules, while hyalinocytes are agranulocytes without these body inclusions. Each type

Table 2.1 Hemocyte types and their functional roles in invertebrates

Phylum	Hemocyte types	Immune role	Pathogen target	Reference
Arthropoda	Granulocytes, Agranulocytes	ROS production, Phagocytosis	Bacteria, Fungi	Eleftherianos et al. (2021)
Mollusca	Granulocytes, Hyalinocytes	AMP secretion, Encapsulation	Vibrio spp.	Chong (2022)
Annelida	Chloragocytes, Amoebocytes	Capsule formation	Parasite larvae	Peters (2021)
Nematoda	Granulocytes	Engulfment & Signaling	Pseudomonas	Hajdú et al. (2024)
Echinodermata	Coelomocytes	Encapsulation, ROS burst	Bacteria	Xue et al. (2024)
Platyhelminthes	Hemocytes	Physical encapsulation	Schistosoma larvae	Al-Khalaifah (2022)
Cnidaria	Amoebocytes	Phagocytosis & Signal Release	Bacteria, Fungi	Snyder et al. (2021)
Porifera	Archaeocytes	Microbe trapping	Bacteria	Ereskovsky et al. (2025)
Rotifera	Hemocytes	Aggregation & Trapping	Protozoa	Raza et al. (2024)
Bryozoa	Coelomocytes	Immune signaling	Bacteria	Queiroz (2020)

of granulocyte contains its immune factors made of enzyme and protein antimicrobials in small granule compartments. The body uses its natural immune defense system to fight pathogens during both harmful substance handling and contaminated object intake. Agranulocytes sense pathogens by identifying them in their surroundings but do not use cytoplasmic granules to fight invaders like other cell types (Knowles et al., 2023). This study examines all known hemocyte varieties in the major and minor animal groups to explain how these animals fight pathogens through their immune cells (Table 2.1).

2.2.1   Phylum Arthropoda (Insects and Crustaceans)

Arthropods depend on their hemocytes as primary immune protectors for both insects and crustaceans. Animals in this group possess both granulocyte and agranulocyte immune cells that function independently to guard the body. In arthropod cells granulocytes perform most immune functions because they kill pathogens directly with toxic elements. In these cells granulocytes store large amounts of enzymes and antimicrobial peptides accessed for fighting pathogens. Agranulocytes own the capacity to find invasive microorganisms and trigger the body's immunity response. *D. melanogaster* inside class Insecta features both granulocyte and agranulocyte cell types among its complete set of hemocytes. Granulocytes defend against microorganisms by releasing the built-up antimicrobial substances present inside these cells (Meister & Lagueux, 2003). Hyalinocytes, or agranulocytes, in

Table 2.2 Cellular immune mechanisms and pathways

Mechanism	Cell type	Pathogen	Molecular mediator	Reference
Phagocytosis	Hemocytes	Bacteria	Lysosomal enzymes, ROS	Liu et al. (2020)
Encapsulation	Granulocytes	Parasites	Melanin, cytokines	Eleftherianos et al. (2021)
ROS Generation	Hemocytes	Fungi, Bacteria	NADPH oxidase	Eleftherianos et al. (2021); Moghadam et al. (2021)
Melanization	Granulocytes	Fungi	Phenoloxidase	Zdybicka-Barabas et al. (2025)
Agglutination	Hemocytes	Vibrio spp.	Lectins	Zhang et al. (2019)
Cell Adhesion	Hemocytes	Fungi	Integrins	Kausar et al. (2022)
Apoptosis	Hemocytes	Viruses	Caspase cascade	Yang et al. (2019)
Nodulation	Hemocytes	Bacteria	Cytokines	Sato (2023)
Extracellular Traps	Granulocytes	Fungi	DNA, Histones	Silva et al. (2021)
Signal Transduction	All hemocytes	Multiple	NF-κB, MAPKs	Wang et al. (2022)

pathogen recognition systems both spot parasites and protect the body by forming a barrier against *Leptopilina boulardi* larvae infection (Kim, 2019). Cells of the immune system rely on hemocytes to produce ROS and AMP defenses, which fight diverse pathogens. *Hemigrapsus sanguineus* uses immune system mechanisms within its granulocytes through enzyme and immune molecule storage in specialized compartments. When bacteria invade their body, these cells act first to capture and destroy microorganisms using the immune response from their disease-fighting mechanisms (Smolowitz, 2021). Granulocytes in crabs recognize pathogenic threats and direct immune responses between cells (Table 2.2).

2.2.2 Phylum Mollusca (Mollusks)

Mollusks use hemocyte cells as their immune system that protects bivalves, gastropods, and cephalopods. Like other arthropods, mollusks depend on both granulocytes and agranulocytes to defend their body against harm. The immune system of mollusks benefits from granulocyte cells that produce killing agents to eliminate invaders while capturing small pathogens. In their basic form hyalinocytes use their immune system skills to identify threats and command other system features to take action. Granulocytes and agranulocytes carry specialized roles in oyster blood when grouped in Mollusca phylum. Oyster granulocytes destroy *Vibrio* infections because of their natural ability to take up disease-causing pathogens. Through PRRs known as pattern recognition receptors, hemocytes can detect foreign pathogens, which prompts an immune response release (Wang et al., 2019). Under the phylum Mollusca, *Aplysia depilans* defends itself by using hemocytes to destroy infections through phagocytosis against bacterial and fungal invaders. The immune system of both granulocytes and agranulocytes defends the host through unique approaches: granulocytes use antimicrobial peptide defense, whereas agranulocytes identify and separate pathogen targets. The immune system of *A. depilans* needs to interact with bacteria (Alesci et al., 2023a), when defending its natural habitat from infections.

2.2.3   Phylum Annelida (Earthworms)

Earthworms defend against infections through specialized immune response methods by two types of hemocytes: granulocytes and agranulocytes. Earthworm granulocytes keep antimicrobial molecules and proteins in cytoplasmic granules to defend the body against infections. When facing intruders, the cells take germs into their body and use their released antimicrobial substances to fight infection. Despite lacking internal granules, hyalinocytes identify microorganisms then launch protection systems by sending cytokines and other signaling molecules. Earthworms in *Lumbricus terrestris* divide their immune cells into granulocytes and agranulocytes. Agranulocytes find foreign invaders and then signal the immune response, but granulocytes work to defend against pathogens using net-like structures and internal enzymes. Earthworm cells make ROS and AMPs to fight bacteria, and agranulocytes make capsules to separate large parasite larvae from their bodies.

2.2.4   Phylum Nematoda (Roundworms)

Caenorhabditis elegans roundworms, together with other nematodes, destroy pathogens using distinct immune cells known as granulocytes and agranulocytes. Granulocytes protect against pathogens through their storage granules in the cytoplasm that fight infections. Granulocytes destroy disease-carrying organisms first by engulfing them and then killing them through phagocytosis (Ermolaeva & Schumacher, 2014). The immune cells without granules in their cytoplasm detect pathogens and launch the defense response through signaling pathways. The immune system of Caenorhabditis elegans cells fights pathogen threats through both granulocytes and agranulocytes. In nematodes, granulocytes eliminate invaders through phagocytosis, while agranulocytes notice threats to trigger defensive mechanisms. These small roundworms need proper immune responses from their immune system to survive infections caused by *Pseudomonas aeruginosa* (Hajdú et al., 2024).

2.2.5 Phylum Echinodermata (Starfish and Sea Urchins)

Echinoderm starfish and sea urchins house both granulocytes and agranulocytes in their different types of hemocytes. Through eating or physical covering of invaders, the immune system works to produce antibacterial proteins and fight off infections. Echinoderm granulocytes protect against infection by ingesting harmful bacteria followed by the release of death-causing substances. Through direct contact with parasites, hyalinocytes of *Asterias rubens* kickstart the immune system of their animal host (Coates et al., 2022). Granulocytes use their antimicrobial peptides and ROS to eliminate pathogenic threats, but agranulocytes mainly detect intruders to trigger immune reactions.

2.2.6 Phylum Platyhelminthes (Flatworms)

The blood cells of flatworm creatures help *Schistosoma mansoni* parasites fight off disease attacks. Hemocytes protect the body by spotting dangerous intruders and then swallowing them along with packaging them into capsules (Frischknecht, 2024). The immune system of flatworms releases capsules from granulocytes to neutralize and eliminate large pathogens, especially *S. mansoni* parasite larvae. Hyalinocytes/agranulocytes work by detecting pathogens and starting their defense systems, while all blood cells protect the host by making anti-pathogenic chemicals and detoxifying foreign organisms. As a parasite, *S. mansoni* manipulates the immune systems of its host to ensure its transport across its new territory. *Schistosoma* hemocytes protect the parasite when they build a physical shield around it, which blocks its movement (Hambrook & Hanington, 2021).

2.2.7 Phylum Cnidaria (Corals and Jellyfish)

Cnidarian animals defend themselves from diseases through different types of immune cells that work together to protect the organism. Cellular parts in body cells of cnidarian animals defend against infections using pathogen destruction and medical protein generation. Immune cells that detect pathogens trigger the entire immune response by releasing immune signals into the body. *Acropora millepora* cells from the phylum *Cnidaria* summon immune responses against dangerous organisms, yet agranulocytes keep these processes working correctly and sending alerts across the immune system. When faced with bacterial and fungal infections, coral immune cells work together to block the infection (Parisi et al., 2020). Every type of invertebrate hemocyte works together with the immune system to defend against many diseases. Invertebrate immune cells exist as granulocytes and agranulocytes, which mainly differ by the presence of pathogen-fighting enzymes in granulocytes or specialized immune signaling and pathogen detection in agranulo-cytes. Hemocytes recognize bacteria and fungi while protecting the body through phagocytosis of invaders and by producing capsules plus enzymes plus giving off reactive oxygen and peptides to fight diseases (Eleftherianos et al., 2021). Scientific research uses immune system responses from diverse living organisms across all major groups of invertebrates to demonstrate how immunity develops across different species. Knowing how hemocytes work lets us understand immune growth and use research to protect plants from insects while improving medical technology.

2.3   Cellular Immune Mechanisms and Pathways

There are several mechanisms involved; below has been provided with detailed information on cell type, pathogen, and molecular mediators (Table 2.2).

2.3.1   Phagocytosis

Invertebrate cells shield their host organism by using phagocytosis as their key natural defense reaction. Hemocytes defend the body by using pathogen receptors to find pathogen-associated molecular patterns (PAMPs) on invaders. The pathogen sensor cells start using their extension tools to create phagosome traps that hold their targets. When the pathogen enters the phagosome, it gets destroyed through enzymes and AMPs combined with ROS. Hemocytes in *D. melanogaster* specimens can identify pathogens through their Toll-like receptors, which they embed into *Escherichia coli* and *Staphylococcus aureus* to sense danger. As soon as pathogens enter the body, the immune system activates different defense lines that strengthen phagocytic cells' performance and enable them to create AMPs and ROS, which destroy the invader (Leclerc & Reichhart, 2004). Hemocytes in Crassostrea *gigas* (Mollusca) kill *Vibrio* species through effective phagocytosis techniques. Through their PRRs, oysters identify bacteria particles. Then their immune cells surround and eliminate the pathogens using AMPs and ROS. A recent research shows that oysters depend on this immune response because their marine habitat faces many pathogen threats (Jiang et al., 2018).

2.3.2   Encapsulation

Blood cells emerge simultaneously to form an outer barrier around large pathogens when single cells cannot take them for removal through encapsulation. Through encapsulation defense, *D. melanogaster* targets this pathogen for removal since the protective barrier holds the harmful invader in a secure area. The immune cells come together in multilayers to entrap invaders through capsule formation before they can penetrate host tissues (Lemaitre & Hoffmann, 2007). *L. terrestris* earthworms from Annelida react to invading parasites by enclosing and containing them. Hemocytes control infection through a physical defense method by enclosing parasitic larvae and microbial invaders to prevent the spread of disease. Because earthworms live in many pathogen-rich areas, their wellness depends on this basic immune cell reaction (Gupta & Yadav, 2016).

2.3.3 Reactive Oxygen Species (ROS) Production

ROS molecules fight infections by being extremely reactive in their task of attacking pathogens. Hemocytes produce ROS when they find pathogens and destroy them using oxidative damage. When harmful substances enter, NADPH oxidase enzymes activate to produce ROS molecules consisting of superoxide radicals and hydrogen peroxide (Hernandez et al., 2022). The immune cells of *C. gigas* in Mollusca produce ROS to defend the organism against Vibrio bacterial infection. Hemocytes in the body generate ROS to kill dangerous organisms and prevent their spread. ROS create the first line of defense against bacterial infection in marine mollusk organisms (Zannella et al., 2017). The infection of *P. aeruginosa* and *E. coli* into *D. melanogaster* leads to the production of reactive oxygen species (ROS) as a defense. ROS removes harmful bacteria by damaging both their outside and inside surfaces, plus their cell walls. Drosophila relies on its ROS killing system as its top defense against diseases (Ramond et al., 2021). *A. rubens* starfish as a member of Echinodermata, protects itself from pests using reactive oxygen species. In response to invading germs, the immune cells of starfish create ROS as their main defense mechanism. Starfish naturally defend itself against disease through the ability of ROS to eliminate harmful organisms (Qu et al., 2024).

2.3.4 Melanization

During an infection response invertebrates activate their immune system to generate melanin. Underneath, dehydroxy phenylalanine transforms into black melanin, which encloses harmful organisms during infection. During infection, the body creates melanin, which forms around the invader and uses oxidation to kill the pathogen (Fig. 2.1). The arthropod Tenebrio molitor mealworm beetle defends itself from bacterial and fungal invaders through melanization defense. To defend against disease pathogens, the body forms melanized walls around infected regions (Vigneron et al., 2019). Scientists studying potential ability of *Drosophila* realized that *D. melanogaster* primarily battles *Aspergillus* fungal infections through its melanization defense mechanism (Mpamhanga & Kounatidis, 2024). Melanin formation

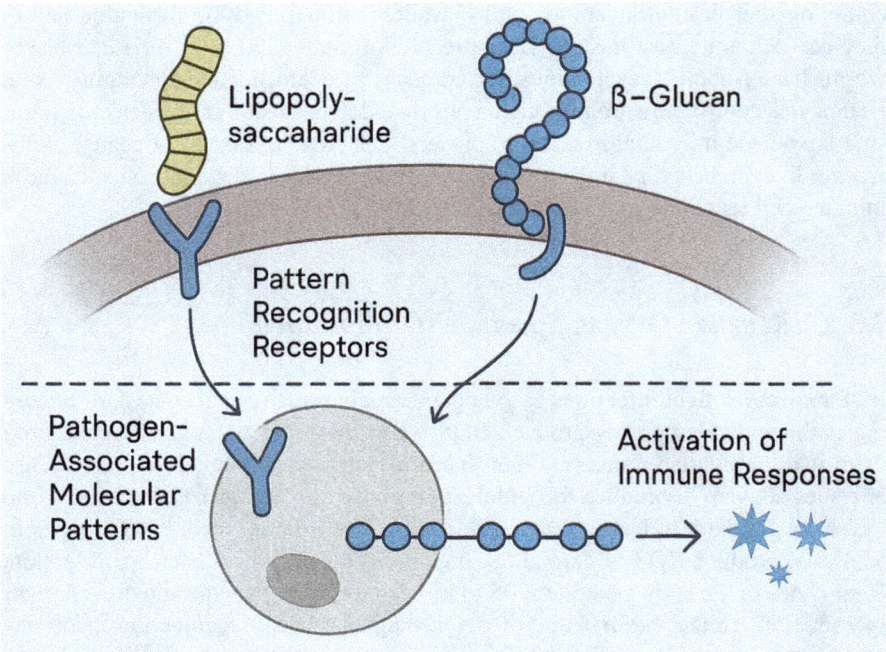

Fig. 2.1 The figure depicts the recognition of pathogen-associated molecular patterns (PAMPs) by pattern recognition receptors (PRRs) and shows downstream activation of immune responses upon binding of PRRs to microbial ligands such as lipopolysaccharides or β-glucans

around infections creates barriers that prevent pathogens from spreading while making the immune system stronger by stopping pathogen growth.

2.3.5 Agglutination

The host immune cells come together to eliminate pathogens before they can spread deeper into the body. Hemocytes neutralize invading viruses and bacteria by linking them together using their ability to clump. When immune cells collect within reach of pathogens they make it simpler for the immune system to eliminate infected areas from the body. Scientists prove that hemocytes from Pacific oysters belonging to *C. gigas* (Mollusca) group to fight off infections. A recent study showed that Mollusca *C. gigas* cells gather together their hemocytes cells to catch and kill *Vibrio* bacteria (Chong, 2022). The blood cells in *Aedes aegypti* mosquitoes from Arthropoda Phylum help fight dengue viruses by gathering them and taking them out of the body. The joint action of mosquitoes prevents foreign viruses from entering their bodies, so they cannot transmit diseases to humans (Cheng et al., 2022).

2.3.6 Cell Adhesion and Entrapment

The immune system cells apply direct pressure to both pathogens and infected tissues to block their movement across the body. Hemocytes in the blood target and attach to pathogens both near and within proper tissue locations to prevent and eliminate invading microorganisms. Hemocytes from *Galleria mellonella* (Arthropoda) bind fungal hyphae during defense with cell adhesion to confine them within protective immune cell clusters (Ding et al., 2020). The immune response puts fungal invaders under control to slow their movements and limit their ability to damage the organism. Earthworms, especially *L. terrestris* from Annelida, use containment methods of their natural defenses to defend themselves from parasites based on their immune mechanisms. When entering, larvae get covered by immune cells; these cells stop the larval development and isolate them to be destroyed (Alesci et al., 2023b). Experts have shown that entrapment stops parasitic diseases from damaging earthworms beyond their current state.

2.4 Surveillance and Monitoring of Cellular-Mediated Immune Responses in Invertebrates

Invertebrate bodies detect and eliminate different pathogens through their immune system functions. Hemocytes serve as immune cells in invertebrates by utilizing cellular defense mechanisms such as detecting pathogens, engulfing them, surrounding them with material, and making AMPs and ROS (Table 2.3). Monitoring the immune system of invertebrates helps scientists find how these natural protectors handle infections. Scientists commonly measure cellular-mediated immune responses in invertebrates through methods that include flow cytometry, microscopy, gene expression methods and more. This section explains immune monitoring in every invertebrate phylum through different measurement techniques.

2.4.1 Flow Cytometry

Flow cytometry serves as one of the primary methods for studying immune cell counts in invertebrate organisms and their immune system reaction. Research shows cells through flow cytometry helps observe if pathogens affect how small or dense hemocytes act and what proteins they produce on their surfaces (Fig. 2.2). Researchers use this technique to discover and count immune cells while observing their activation patterns during infections. Flow cytometry measurements show how *D. melanogaster* hemocytes become active during bacterial infections with *E. coli*. The research team saw immune cells transforming in their size and surface marker activity as they recorded these responses during infected periods

Table 2.3 Quantitative methods for monitoring cellular immunity

Method	Target parameter	Purpose	Model species	Reference
Flow Cytometry	Hemocyte count	Immune profiling	Shrimp	Xian et al. (2021)
ROS Assay	ROS levels	Oxidative defense	Oyster	Xing et al. (2023)
Phagocytosis Assay	Engulfment index	Pathogen uptake	Drosophila	Liégeois et al. (2020)
Encapsulation Index	Capsule formation	Parasite defense	Earthworm	Prochazkova et al. (2019)
Hemocyte Spreading	Cell activation	Adhesion & response	Mollusks	Lv et al. (2022)
Trypan Blue Exclusion	Hemocyte viability	Cell death detection	Crustaceans	Kang et al. (2025)
RT-qPCR	Gene expression	Immune gene upregulation	Drosophila	Huang et al. (2024)
Fluorescent Microscopy	ROS production	Live cell imaging	Starfish	Andrade et al. (2021)
Immunoblotting	PO pathway proteins	Pathway validation	Crabs	Yang et al. (2022)
Nodulation Score	Nodule count	Infection load	Insects	Sato (2023)

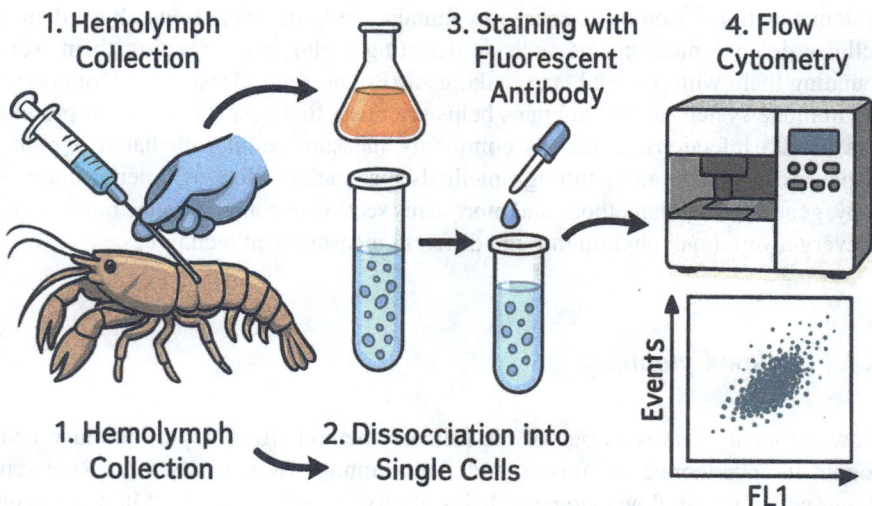

Fig. 2.2 The figure presents flow cytometry data characterizing hemocyte populations based on size, granularity, and marker expression and demonstrates the identification and quantification of distinct hemocyte subsets involved in immune defense

(Troha & Buchon, 2019). City Science Center uses flow cytometry at *H. sanguineus* to analyze the immune system of Crustacean related to *Vibrio* infection (Epifanio, 2013). Recent research shows how infected hemocytes become active and fight microbes along with their defense peptides.

2.4.2   Confocal and Fluorescence Microscopy

Hemocytes and pathogens interact under direct observation using confocal and fluorescence microscopy. Researchers use live Imaging Methods to see immune process activities including how cells engulf germs and produce ROS. Special fluorescent labels help scientists watch how immune cells respond to pathogens while the markers show the pathogens leaving the organism (Fig. 2.3). Scientists use confocal microscopy to study how *P. aeruginosa* and *E. coli* affect hemocytes when infecting the flies *D. melanogaster*. The scientists can watch bacteria intake by immune cells in living specimens while tracking how immune cells encompass parasitic larvae to better understand the mechanisms driving pathogen fighting (Fauvarque, 2014). Studies show that Vibrio species interactions with oyster blood cells using this technique in *C. gigas*. This procedure helps scientists observe how hemocytes defend against pathogens through pathogen detection and phagocytic activity plus immune aggregate formation (Jiang et al., 2018).

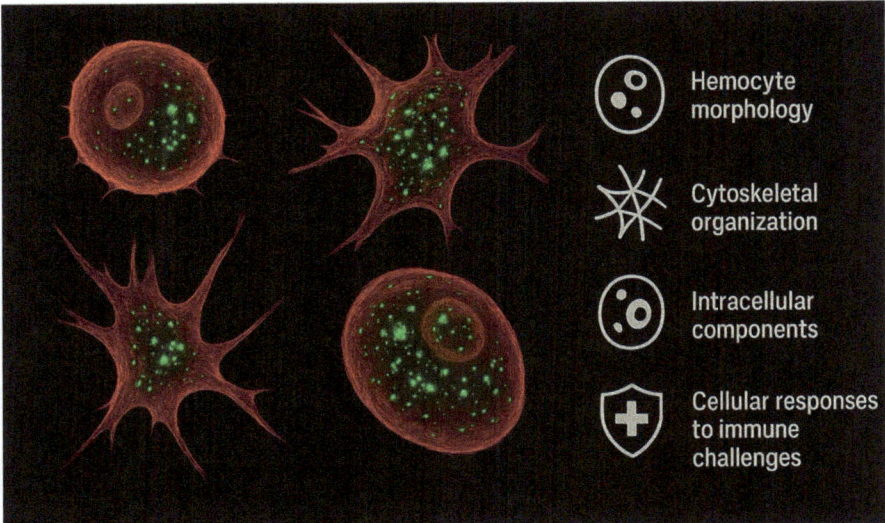

Fig. 2.3 Confocal laser scanning microscopy images highlighting hemocyte morphology, cytoskeletal organization, and intracellular components. The figure illustrates cellular responses to immune challenges at high resolution

2.4.3   RNA Sequencing (RNA-Seq)

RNA sequencing offers researchers a superior method to identify changes in gene expression that happen when cells encounter pathogens. Through RNA profiling studies, scientists find immune genes connected to phagocytosis and ROS creation, plus the synthesis of antimicrobial peptides (Table 2.4). The RNA-Seq test shows us how many immune genes become more active and how many become less active when the host fights an infection. Research using RNA-Seq examines *Drosophila* immune reactions to *P. aeruginosa* infections through the phylum Arthropoda. By using this technique scientists found which genes activate AMPs such as cecropins and defensins (Castillo et al., 2015) (Fig. 2.4). Scientists applied RNA-Seq tests to measure *C. gigas's* genetic response when *Vibrio* species infected the Pacific oyster from the Phylum Mollusca. The results show that immune genes that recognize pathogens and produce ROS plus AMPs are activated (Zhang et al., 2024) (Fig. 2.5).

2.4.4   Immunohistochemistry

People commonly use immunohistochemistry to study how invertebrates fight infections. Researchers use tissue staining with immune-related protein antibodies to examine how cells in the immune system protect hosts from infection by *Vibrio* species (Coates et al., 2022). This technique shows the number and activation state

Table 2.4 Immune cell subtypes and morphology across invertebrates

Phylum	Subtype	Morphological trait	Function	Reference
Arthropoda	Plasmatocytes	Round, granule-less	Phagocytosis	Ravindranath (1978)
	Granulocytes	Granule-rich, elongated	ROS, encapsulation	Baxter et al. (2017)
Mollusca	Hyalinocytes	Transparent cytoplasm	Signaling, migration	Ma et al. (2024)
	Granulocytes	Dense granules	AMP secretion	Canesi et al. (2022)
Annelida	Chloragocytes	Vacuolated	Enzyme synthesis	Fischer (1993)
Echinodermata	Phagocytes	Pseudopodia	Microbe engulfment	Smith et al. (2018)
Platyhelminthes	Hemocytes	Irregular shapes	Capsule formation	Adell et al. (2015)
Nematoda	Granulocytes	Cytoplasmic granules	Pathogen digestion	Van Meulder et al. (2013)
Cnidaria	Amoebocytes	Amoeboid shape	Immune response	Parisi et al. (2020)
Porifera	Archaeocytes	Spherical nucleus	Nutrient & microbe control	Ereskovsky et al. (2025)

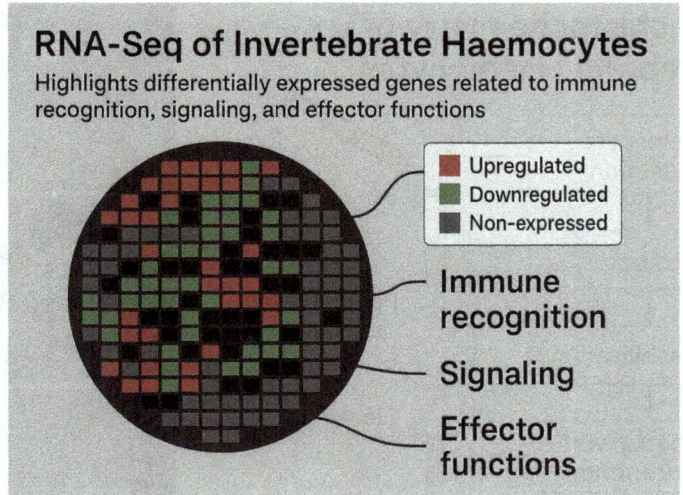

Fig. 2.4 The figure shows transcriptomic profiling results from hemocytes under various immune stimuli and highlights differentially expressed genes related to immune recognition, signaling, and effector functions

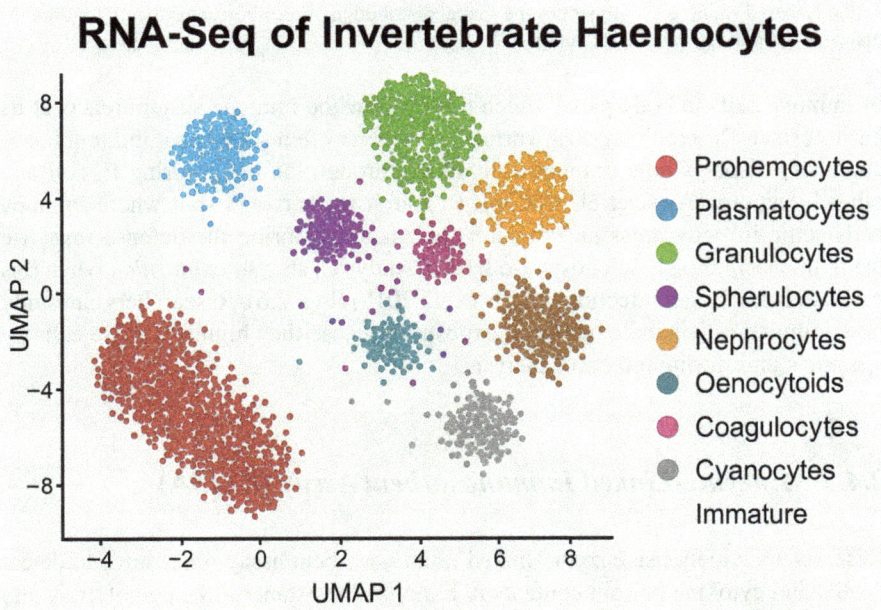

Fig. 2.5 RNA-Seq of invertebrate hemocytes showing several subtypes

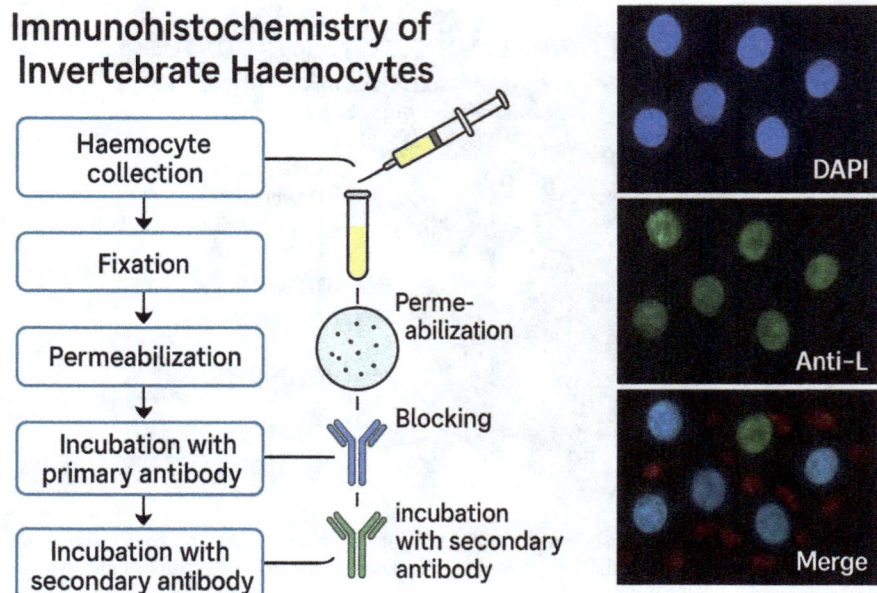

Fig. 2.6 The figure displays localization of specific immune proteins within hemocytes using antibody-based staining and illustrates the spatial distribution of receptors, enzymes, or antimicrobial peptides during immune activation

of immune cells in body parts, which reveals how the immune system reacts at its basic format. Research on earthworms *L. terrestris* often shows that immunohistochemistry displays how immune cells move through the body during E. coli and other infections (Alesciet al., 2023b). Research workers can spot where immune cells settle in body areas and monitor their activity during the defense response using this technology. Scientists use IHC to study sea slug *A. californica* when this mollusk faces fungal infections (Avila et al., 2018) (Fig. 2.6). Researchers can study how immune cells handle pathogen exposure because they highlight these cells by specific stains during infection analysis.

2.4.5   *Enzyme-Linked Immunosorbent Assay (ELISA)*

Researchers widely use enzyme-linked immunosorbent assay procedures to detect AMPs and cytokine protein contents in biological substances like hemolymph and tissue samples. Our method helps researchers determine the amount of immune-related substances and shows the immune system's effectiveness against pathogens (Fig. 2.7). A recent study used ELISA tests to measure defensin amounts in *C. gigas* as part of studies on pathogen defense. The team measures AMP concentrations both before and after infection to evaluate how well the immune system defends the

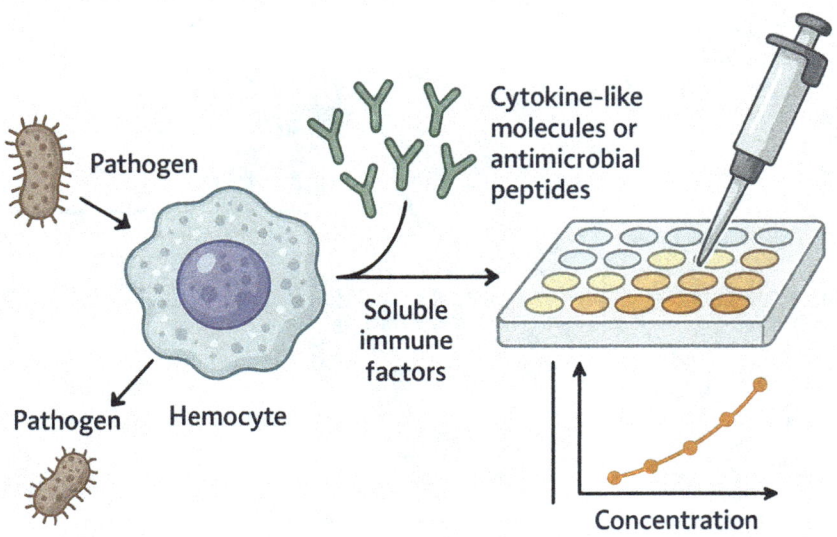

Fig. 2.7 Quantitative measurement of soluble immune factors produced by hemocytes using enzyme-linked immunosorbent assay. It shows concentrations of cytokine-like molecules or antimicrobial peptides in response to pathogens

oyster against pathogens (Zannella et al., 2017). A research study on *D. melanogaster shows that* that utilizing ELISA to analyze cecropin and attacin production when the organism fights *E. coli* infection. These testing procedures reveal how well the immune system activates immune proteins to fight off microbes (Carboni et al., 2022).

2.4.6   *Proteomics*

The scientific field of proteomics evaluates many proteins, especially those defending against infection. Research teams use protein detection to reveal how cells defend invertebrate lifeforms against invaders at the molecular level. Proteomics shows changes in immune protein production, such as pathogen detection proteins and phagocytic cells, along with ROS regulators. Scientists studied the bacterial immunity response proteins in *D. melanogaster* flies using proteomic technologies. The scientific team analyzed protein samples to discover significant immune system protectors and their participation in testing for pathogens, plus switching on immune protection systems (Levy et al., 2004). In *C. gigas,* proteomics research shows which proteins are at work when this oyster fights *Vibrio* bacteria. With this method scientists gained important information on immune cell proteins and their antibacterial fighting functions (Jiang et al., 2018). The immune response of invertebrate

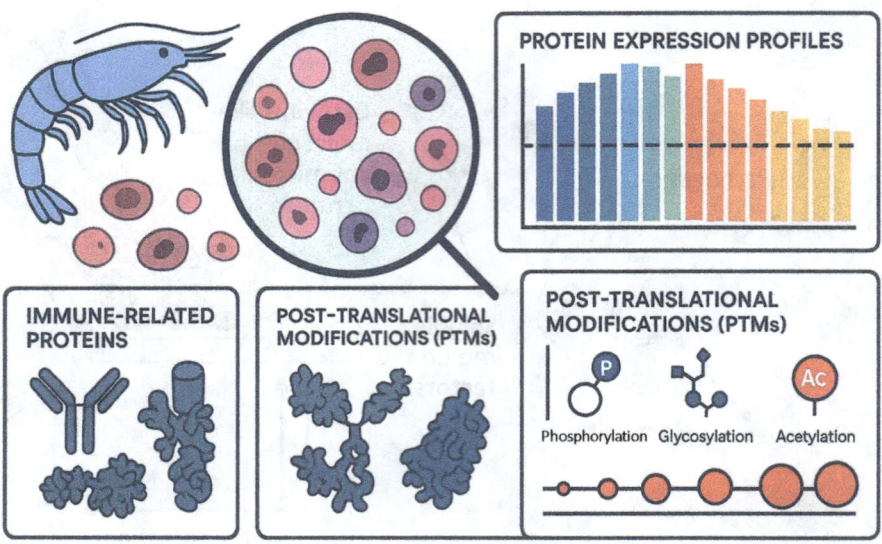

Fig. 2.8 Comprehensive analysis of protein expression profiles in hemocytes. The figure highlights immune-related proteins, post-translational modifications, and pathway enrichments contributing to host defense

cells needs regular watching to reveal how pathogens affect defense against pathogens. Scientific tools help researchers see changes in immune cell actions and pathogen interactions by using flow cytometry microscopy, RNA sequencing immunohistochemistry, ELISA, and proteomics tests. Experts use various methods to study immune response in different invertebrates, including *D. melanogaster* arthropods, *C. gigas* molluscs, and *L. terrestris* annelids (Söderhäll, 2024). Our knowledge of immune reactions helps both basic scientists and enables us to create solutions that fight pests with disease treatments and new antibiotics (Fig. 2.8).

2.5   Conclusion

The surveillance and monitoring of cellular-mediated immune responses in invertebrates is critical to understanding their complex defense mechanisms, particularly because these organisms lack an adaptive immune system. Invertebrates rely heavily on their innate immune responses, which are mediated primarily through hemocytes, specialized immune cells that patrol the hemolymph for pathogens. These hemocytes are categorized into granulocytes and agranulocytes, each playing distinct roles in pathogen recognition, pathogen elimination, and immune signaling. Granulocytes, equipped with enzymes and antimicrobial peptides (AMPs) stored in

cytoplasmic granules, are crucial in directly eliminating pathogens through phagocytosis, encapsulation, and the production of reactive oxygen species (ROS). On the
other hand, agranulocytes are essential for detecting pathogens and triggering
immune responses by signaling and initiating further defense mechanisms. Together,
these two types of hemocytes form the backbone of invertebrate immune defense.
Examples from various invertebrate phyla, including arthropods, mollusks, annelids, and nematodes, highlight the diverse strategies these organisms employ to protect themselves from pathogens. In *D. melanogaster*, granulocytes and agranulocytes
coordinate efforts to identify and neutralize bacterial threats, with the immune
response involving both direct pathogen destruction and signaling pathways to
recruit other immune components. Similarly, mollusks like *C. gigas* use granulocytes to clear infections from bacteria like *Vibrio*, while agranulocytes help in
pathogen recognition and initiating immune signaling pathways. In annelids such as
L. terrestris, granulocytes utilize antimicrobial molecules stored in their granules to
protect against bacterial threats, and agranulocytes contribute by producing encapsulation responses against parasites.

The ability to study these immune responses through advanced techniques like
flow cytometry, microscopy, gene expression analysis, and proteomics offers a powerful means to explore invertebrate immunity at a molecular level. These tools
enable researchers to visualize the interactions between immune cells and pathogens in real time, measure the levels of immune proteins, and identify the key
molecular pathways involved in immune activation. Flow cytometry, for instance,
allows the identification and quantification of immune cell populations and their
activation states, while RNA sequencing provides a detailed picture of the gene
expression changes that occur during infection. Proteomics, on the other hand, identifies and quantifies proteins involved in immune responses, providing insight into
how immune cells work to combat pathogens. As the tools and techniques used to
monitor and study cellular-mediated immune responses in invertebrates continue to
evolve, they offer valuable opportunities for improving pest control strategies,
enhancing aquaculture health, and developing novel disease prevention methods.
The knowledge gained from invertebrate immunity research not only advances our
understanding of innate immunity but also has broader applications in medicine,
agriculture, and environmental science. Future research in this field will continue to
illuminate the complex immune systems of invertebrates, offering innovative solutions to address the global challenges posed by infectious diseases and pest management. Thus, the exploration of invertebrate immune systems through
cellular-mediated immune responses continues to provide essential insights into the
evolution and regulation of immune defenses across species, paving the way for
future breakthroughs in immune therapeutics and environmental health management.

References

Adell, T., Martín-Durán, J. M., Saló, E., & Cebrià, F. (2015). Platyhelminthes. In *Evolutionary developmental biology of invertebrates 2: Lophotrochozoa (Spiralia)* (pp. 21–40). Springer Vienna.

Alesci, A., Capillo, G., Fumia, A., Albano, M., Messina, E., Spanò, N., et al. (2023a). Coelomocytes of the Oligochaeta earthworm Lumbricus terrestris (Linnaeus, 1758) as evolutionary key of defense: A morphological study. *Zoological Letters, 9*(1), 5.

Alesci, A., Fumia, A., Albano, M., Messina, E., D'Angelo, R., Mangano, A., et al. (2023b). Investigating the internal system of defense of Gastropoda Aplysia depilans (Gmelin, 1791): Focus on hemocytes. *Fish & Shellfish Immunology, 137*, 108791.

Al-Khalaifah, H. (2022). Cellular and humoral immune response between snail hosts and their parasites. *Frontiers in Immunology, 13*, 981314.

Andrade, C., Oliveira, B., Guatelli, S., Martinez, P., Simões, B., Bispo, C., et al. (2021). Characterization of coelomic fluid cell types in the starfish Marthasterias glacialis using a flow cytometry/imaging combined approach. *Frontiers in Immunology, 12*, 641664.

Avila, C., Núñez-Pons, L., & Moles, J. (2018). From the tropics to the poles: Chemical defense strategies in sea slugs (Mollusca: Heterobranchia). In *Chemical ecology* (pp. 71–163). CRC Press.

Baxter, R. H., Contet, A., & Krueger, K. (2017). Arthropod innate immune systems and vector-borne diseases. *Biochemistry, 56*(7), 907–918.

Betancourt, J. L., Rodríguez-Ramos, T., & Dixon, B. (2024a). Pattern recognition receptors in Crustacea: Immunological roles under environmental stress. *Frontiers in Immunology, 15*, 1474512.

Canesi, L., Auguste, M., Balbi, T., & Prochazkova, P. (2022). Soluble mediators of innate immunity in annelids and bivalve mollusks: A mini-review. *Frontiers in Immunology, 13*, 1051155.

Carboni, A. L., Hanson, M. A., Lindsay, S. A., Wasserman, S. A., & Lemaitre, B. (2022). Cecropins contribute to Drosophila host defense against a subset of fungal and Gram-negative bacterial infection. *Genetics, 220*(1), iyab188.

Castillo, J. C., Creasy, T., Kumari, P., Shetty, A., Shokal, U., Tallon, L. J., & Eleftherianos, I. (2015). Drosophila anti-nematode and antibacterial immune regulators revealed by RNA-Seq. *BMC Genomics, 16*(1), 519.

Cheng, L., Liu, W. L., Su, M. P., Huang, S. C., Wang, J. R., & Chen, C. H. (2022). Prohemocytes are the main cells infected by dengue virus in Aedes aegypti and Aedes albopictus. *Parasites & Vectors, 15*(1), 137.

Chong, R. S. M. (2022). Molluscan immunology. In *Aquaculture pathophysiology* (pp. 383–392). Academic Press.

Coates, C. J., Rowley, A. F., Smith, L. C., & Whitten, M. M. (2022). Host defences of invertebrates to pathogens and parasites. In *Invertebrate pathology* (Vol. 1). Oxford University Press.

Ding, J. L., Hou, J., Feng, M. G., & Ying, S. H. (2020). Transcriptomic analyses reveal comprehensive responses of insect hemocytes to mycopathogen Beauveria bassiana, and fungal virulence-related cell wall protein assists pathogen to evade host cellular defense. *Virulence, 11*(1), 1352–1365.

Eleftherianos, I., Heryanto, C., Bassal, T., Zhang, W., Tettamanti, G., & Mohamed, A. (2021). Haemocyte-mediated immunity in insects: Cells, processes and associated components in the fight against pathogens and parasites. *Immunology, 164*(3), 401–432.

Epifanio, C. E. (2013). Invasion biology of the Asian shore crab Hemigrapsus sanguineus: A review. *Journal of Experimental Marine Biology and Ecology, 441*, 33–49.

Ereskovsky, A., Melnikov, N. P., & Lavrov, A. (2025). Archaeocytes in sponges: Simple cells of complicated fate. *Biological Reviews, 100*(2), 790–814.

Ermolaeva, M. A., & Schumacher, B. (2014). Insights from the worm: The C. elegans model for innate immunity. In *Seminars in immunology* (Vol. 26, No. 4, pp. 303–309). Academic Press.

Fauvarque, M. O. (2014). Small flies to tackle big questions: Assaying complex bacterial virulence mechanisms using D rosophila melanogaster. *Cellular Microbiology, 16*(6), 824–833.

Fischer, E. (1993). The myelo-erythroid nature of the chloragogenous-like tissues of the annelids. *Comparative Biochemistry and Physiology Part A: Physiology, 106*(3), 449–453.

Frischknecht, F. (2024). *Parasites*. Springer Fachmedien Wiesbaden.

Gupta, S., & Yadav, S. (2016). Immuno-defense strategy in earthworms: A review article. *International Journal of Current Microbiology and Applied Sciences, 5*, 1022–1035.

Hajdú, G., Szathmári, C., & Sőti, C. (2024). Modeling host–pathogen interactions in C. elegans: Lessons learned from Pseudomonas aeruginosa infection. *International Journal of Molecular Sciences, 25*(13), 7034.

Hambrook, J. R., & Hanington, P. C. (2021). Immune evasion strategies of schistosomes. *Frontiers in Immunology, 11*, 624178.

Hernandez, E. P., Anisuzzaman, A., Kawada, H., Kwofie, K. D., Ladzekpo, D., et al. (2022). Ambivalent roles of oxidative stress in triangular relationships among arthropod vectors, pathogens and hosts. *Antioxidants, 11*(7), 1254.

Huang, Y., Pang, Y., Xu, Y., Liu, L., & Zhou, H. (2024). The identification of regulatory ceRNA network involved in Drosophila Toll immune responses. *Developmental & Comparative Immunology, 151*, 105105.

Jiang, S., Qiu, L., Wang, L., Jia, Z., Lv, Z., Wang, M., et al. (2018). Transcriptomic and quantitative proteomic analyses provide insights into the phagocytic killing of hemocytes in the oyster Crassostrea gigas. *Frontiers in Immunology, 9*, 1280.

Kang, F., Xiao, B., Fan, T., Li, Q., Liu, M., He, J., & Li, C. (2025). Subpopulation-specific apoptotic responses of Hemocytes to Decapod Iridescent Virus 1 in Macrobrachium rosenbergii. *Fish & Shellfish Immunology, 167*, 110685.

Kausar, S., Abbas, M. N., Gul, I., Liu, Y., Tang, B. P., Maqsood, I., et al. (2022). Integrins in the immunity of insects: A review. *Frontiers in Immunology, 13*, 906294.

Kim, C. (2019). *Drosophila immune response to the endoparasitoid wasp Leptopilina boulardi: characterization of a resistance reaction* (Doctoral dissertation, COMUE Université Côte d'Azur (2015–2019)).

Knowles, S., Dennis, M., McElwain, A., Leis, E., & Richard, J. (2023). Pathology and infectious agents of unionid mussels: A primer for pathologists in disease surveillance and investigation of mortality events. *Veterinary Pathology, 60*(5), 510–528.

Leclerc, V., & Reichhart, J. M. (2004). The immune response of Drosophila melanogaster. *Immunological Reviews, 198*(1), 59–71.

Lemaitre, B., & Hoffmann, J. (2007). The host defense of Drosophila melanogaster. *Annual Review of Immunology, 25*(1), 697–743.

Levy, F., Bulet, P., & Ehret-Sabatier, L. (2004). Proteomic analysis of the systemic immune response of Drosophila. *Molecular & Cellular Proteomics, 3*(2), 156–166.

Liégeois, S., Wang, W., & Ferrandon, D. (2020). Methods to quantify in vivo phagocytic uptake and Opsonization of live or killed microbes in Drosophila melanogaster. In *Immunity in insects* (pp. 79–95). Springer US.

Liu, S., Zheng, S. C., Li, Y. L., Li, J., & Liu, H. P. (2020). Hemocyte-mediated phagocytosis in crustaceans. *Frontiers in Immunology, 11*, 268.

Lv, Z., Qiu, L., Wang, W., Liu, Z., Liu, Q., Wang, L., & Song, L. (2022). RGD-labeled hemocytes with high migration activity display a potential immunomodulatory role in the pacific oyster Crassostrea gigas. *Frontiers in Immunology, 13*, 914899.

Ma, Z., Wu, Y., Zhang, Y., Zhang, W., Jiang, M., Shen, X., et al. (2024). Morphologic, cytometric, quantitative transcriptomic and functional characterisation provide insights into the haemocyte immune responses of Pacific abalone (Haliotis discus hannai). *Frontiers in Immunology, 15*, 1376911.

Meister, M., & Lagueux, M. (2003). Drosophila blood cells. *Cellular Microbiology, 5*(9), 573–580.

Moghadam, Z. M., Henneke, P., & Kolter, J. (2021). From flies to men: ROS and the NADPH oxidase in phagocytes. *Frontiers in Cell and Developmental Biology, 9*, 628991.

Mpamhanga, C. D., & Kounatidis, I. (2024). The utility of Drosophila melanogaster as a fungal infection model. *Frontiers in Immunology, 15*, 1349027.

Parisi, M. G., Parrinello, D., Stabili, L., & Cammarata, M. (2020). Cnidarian immunity and the repertoire of defense mechanisms in anthozoans. *Biology, 9*(9), 283.

Patnaik, B. B., Baliarsingh, S., Sarkar, A., Hameed, A. S., Lee, Y. S., Jo, Y. H., et al. (2024). The role of pattern recognition receptors in crustacean innate immunity. *Reviews in Aquaculture, 16*(1), 190–233.

Peters, E. C. (2021). Diseases of other invertebrate phyla: Porifera, cnidaria, ctenophora, annelida, echinodermata. In *Pathobiology of marine and estuarine organisms* (pp. 393–449). CRC Press.

Prochazkova, P., Roubalova, R., Skanta, F., Dvorak, J., Pacheco, N. I. N., Kolarik, M., & Bilej, M. (2019). Developmental and immune role of a novel multiple cysteine cluster TLR from Eisenia andrei earthworms. *Frontiers in Immunology, 10*, 1277.

Qu, L., Sun, Y., Zhao, C., Elphick, M. R., & Wang, Q. (2024). Research progress on starfish outbreaks and their prevention and utilization: Lessons from northern China. *Biology, 13*(7), 537.

Queiroz, V. (2020). An unprecedented association of an encrusting bryozoan on the test of a live sea urchin: Epibiotic relationship and physiological responses. *Marine Biodiversity, 50*(5), 86.

Ramond, E., Jamet, A., Ding, X., Euphrasie, D., Bouvier, C., Lallemant, L., et al. (2021). Reactive oxygen speciesdependent innate immune mechanisms control methicillin-resistant Staphylococcus aureus virulence in the Drosophila larval model. *MBio, 12*(3), 10–1128.

Ravindranath, M. H. (1978). The individuality of plasmatocytes and granular hemocytes of arthropods—A review. *Developmental & Comparative Immunology, 2*(4), 581–594.

Raza, B., Zheng, Z., & Yang, W. (2024). A review on biofloc system technology, history, types, and future economical perceptions in aquaculture. *Animals, 14*(10), 1489.

Sato, R. (2023). Mechanisms and roles of the first stage of nodule formation in lepidopteran insects. *Journal of Insect Science, 23*(4), 3.

Silva, J. D. C., Thompson-Souza, G. D. A., Barroso, M. V., Neves, J. S., & Figueiredo, R. T. (2021). Neutrophil and eosinophil DNA extracellular trap formation: Lessons from pathogenic fungi. *Frontiers in Microbiology, 12*, 634043.

Smith, L. C., Arizza, V., Barela Hudgell, M. A., Barone, G., Bodnar, A. G., Buckley, K. M., et al. (2018). Echinodermata: The complex immune system in echinoderms. In *Advances in comparative immunology* (pp. 409–501). Springer International Publishing.

Smolowitz, R. (2021). Arthropoda: Decapoda. In *Invertebrate histology* (pp. 277–299).

Snyder, G. A., Eliachar, S., Connelly, M. T., Talice, S., Hadad, U., Gershoni-Yahalom, O., et al. (2021). Functional characterization of Hexacorallia phagocytic cells. *Frontiers in Immunology, 12*, 662803.

Söderhäll, K. (2024). Invertebrate immunology–some thoughts about past and future research. *Developmental & Comparative Immunology, 161*, 105256.

Troha, K., & Buchon, N. (2019). Methods for the study of innate immunity in Drosophila melanogaster. *Wiley Interdisciplinary Reviews: Developmental Biology, 8*(5), e344.

Van Meulder, F., Van Coppernolle, S., Borloo, J., Rinaldi, M., Li, R. W., Chiers, K., et al. (2013). Granule exocytosis of granulysin and granzyme B as a potential key mechanism in vaccine-induced immunity in cattle against the nematode Ostertagia ostertagi. *Infection and Immunity, 81*(5), 1798–1809.

Vigneron, A., Jehan, C., Rigaud, T., & Moret, Y. (2019). Immune defenses of a beneficial pest: The mealworm beetle, Tenebrio molitor. *Frontiers in Physiology, 10*, 138.

Wang, L., Zhang, H., Wang, M., Zhou, Z., Wang, W., Liu, R., et al. (2019). The transcriptomic expression of pattern recognition receptors: Insight into molecular recognition of various invading pathogens in Oyster Crassostrea gigas. *Developmental & Comparative Immunology, 91*, 1–7.

Wang, S., Li, H., Chen, R., Jiang, X., He, J., & Li, C. (2022). TAK1 confers antibacterial protection through mediating the activation of MAPK and NF-κB pathways in shrimp. *Fish & Shellfish Immunology, 123*, 248–256.

Xian, J. A., Zheng, P. H., Lu, Y. P., Li, J. T., Zhang, Z. L., Zhang, X. X., et al. (2021). Flow cytometric analysis of morphologic and immunological characterisation of the tiger shrimp Penaeus monodon haemocytes. *Aquaculture Reports, 20*, 100748.

Xing, Z., Gao, L., Liu, R., Yang, Q., Li, Q., Wang, L., & Song, L. (2023). The oxidative stress of the Pacific oyster Crassostrea gigas under high-temperature stress. *Aquaculture, 577*, 739998.

Xue, Z., Peng, T., & Wang, W. (2024). Classification and immune function of coelomocytes in echinoderms. *Current Chinese Science, 4*(1), 25–43.

Yang, G., Wang, J., Luo, T., & Zhang, X. (2019). White spot syndrome virus infection activates Caspase 1-mediated cell death in crustacean. *Virology, 528*, 37–47.

Yang, Z., Zhou, J., Zhu, L., Chen, A., & Cheng, Y. (2022). Label-free quantification proteomics analysis reveals acute hyperosmotic responsive proteins in the gills of Chinese mitten crab (Eriocheir sinensis). *Comparative Biochemistry and Physiology Part D: Genomics and Proteomics, 43*, 101009.

Zannella, C., Mosca, F., Mariani, F., Franci, G., Folliero, V., Galdiero, M., et al. (2017). Microbial diseases of bivalve mollusks: Infections, immunology and antimicrobial defense. *Marine Drugs, 15*(6), 182.

Zdybicka-Barabas, A., Stączek, S., Kunat-Budzyńska, M., & Cytryńska, M. (2025). Innate immunity in insects: The lights and shadows of Phenoloxidase system activation. *International Journal of Molecular Sciences, 26*(3), 1320.

Zhang, E., Li, Z., Dong, L., Feng, Y., Sun, G., Xu, X., et al. (2024). Exploration of molecular mechanisms of immunity in the pacific oyster (Crassostrea gigas) in response to Vibrio alginolyticus invasion. *Animals, 14*(11), 1707.

Zhang, J., Zhang, Y., Chen, L., Yang, J., Wei, Q., Yang, B., et al. (2019). Two c-type lectins from Venerupis philippinarum: Possible roles in immune recognition and opsonization. *Fish & Shellfish Immunology, 94*, 230–238.

Chapter 3
Surveillance and Monitoring of Humoral Immune Responses

3.1 Introduction

The defense mechanisms operate through humoral immune proteins, which include pattern recognition proteins (PRPs) as well as antimicrobial peptides (AMPs) and multiple other immune molecules. The proteins play three vital roles in immune detection of pathogens while enabling immune system activation and microbial threat elimination. PRPs operate as fundamental components of the humoral system by identifying pathogen-associated molecular patterns (PAMPs) on microbial surfaces. The detection process activates immune responses that produce AMPs together with reactive oxygen species (ROS) and additional immune molecules, which eliminate pathogens. Different invertebrate species, including shrimp and crayfish, along with mollusks, together with insects, will be examined in this section regarding their humoral immune proteins' roles. The study includes an evaluation of pattern recognition proteins PGBP and PGRP, along with BGBP and BGRP, and Toll receptors throughout invertebrate phyla, along with their activation methods (Table 3.1).

3.1.1 *Peptidoglycan Binding Protein (PGBP)*

Pattern recognition protein Peptidoglycan Binding Protein (PGBP) functions as an essential element that detects peptidoglycan, which constitutes the main substance of bacterial cell walls. The recognition of bacterial pathogens through PGBP leads to activation of immune responses, which finally result in pathogen neutralization. Many crustaceans together with mollusks utilize PGBP to defend themselves against Gram-positive bacteria throughout their organisms. In *Penaeus monodon*, a species of shrimp, PGBP binds to peptidoglycan from the bacterial cell walls of

Table 3.1 Pattern recognition proteins in invertebrates

PRP type	Ligand	Host	Immune role	Reference
PGRP	Peptidoglycan	*Drosophila*	AMP induction	Royet (2004)
PGBP	Peptidoglycan	Shrimp	ROS + AMP activation	Udompetcharaporn et al. (2014)
BGBP	β-glucans	Shrimp	Fungal detection	Singh and Bhardwaj (2023)
BGRP	β-glucans	Oyster	ROS production	Zhao et al. (2023)
TLR	LPS, β-glucan	*Drosophila*	Toll signaling	Hackett (2003)
Lectins	Mannose	Mollusks	Agglutination	Hassan et al. (2020)
Ficolin	GlcNAc	Clam	ROS/AMP induction	Xiang et al. (2014)
Galectin	β-galactosides	Earthworm	Agglutination & opsonization	Cerenius and Söderhäll (2021)
SRs	Lipids, PAMPs	Crustaceans	Endocytosis	Mahla et al. (2013)
FREP	Glycans	Mollusks	Immune memory-like opsonization	Adema (2015)

pathogens like *Vibrio* species. After binding occurs, PGBP triggers immune response activation, which leads to the production of antimicrobial peptides that eliminate pathogens. The shrimp relies on this recognition procedure to defend itself against various types of bacterial infections (Rosilan et al., 2023; Udompetcharaporn et al., 2014). Similar to *C. gigas,* PGBP enables bacterial recognition of *Vibrio* species pathogens. When PGBP interacts with peptidoglycan, the pharmaceutical response starts, which ultimately produces AMPs that help kill bacteria (Lin et al., 2008).

3.1.2 Peptidoglycan Recognition Protein (PGRP)

PGRPs constitute a protein family that identifies peptidoglycan, which exists within Gram-negative and Gram-positive bacterial cell walls. The immune system of invertebrates uses PGRPs to start defense mechanisms that trigger antimicrobial peptide and various immune molecule production pathways. The red swamp crayfish species *P. clarkii* depends on PGRP to detect microbial pathogens, especially *E. coli* and *Vibrio* bacteria. When pathogens are recognized by PGRP, these proteins initiate signals that engage immune pathways to create AMPs which boost the crayfish resistance against infections (Ren et al., 2024). In *Drosophila*, PGRP acts as a sensor for peptidoglycan in bacterial cell walls. PGRP activation by bacterial pathogens starts the IMD pathway, which produces attacins and cecropin AMPs to stop bacterial infections (Liegeois & Ferrandon, 2022; Royet, 2004).

3.1.3 Beta-Glucan Binding Protein (BGBP)

The immune system of crustaceans detects beta-glucans using Beta-glucan Binding Protein (BGBP) as these carbohydrates exist within the defensive membranes surrounding fungi and specific bacterial cells. BGBP functions as an essential element for detection of fungal infection, which triggers immune mechanisms that aid in fighting pathogenic fungi. The Beta-glucan Binding Protein in shrimp species *L. vannamei* detects fungal pathogen beta-glucan molecules. Upon beta-glucans binding to BGBP a sequence of immune responses starts, which proceeds to generate AMPs and ROS production to destroy the invading fungi (Singh & Bhardwaj, 2023). Drosophila utilize BGBP to detect the fungal pathogen Aspergillus and similar genera by their beta-glucan structures. The immune system becomes active when BGBP binds to fungal pathogens as it sets off the production of specific antifungal AMPs that protect the fly from infections (Zhou et al., 2024).

3.1.4 Beta-Glucan Recognition Protein (BGRP)

The beta-glucan Recognition Protein (BGRP) operates like BGBP for fungal pathogen beta-glucan recognition purposes. The binding of BGRP to fungal cell walls activates immune responses through which the invaders get neutralized. In the mollusk *C. gigas*, BGRP binds to beta-glucans from fungal pathogens like *Candida* species (Zhao et al., 2023). When BGRP detects fungal components, it activates an immune response that ends in the production of AMPs and ROS, leading to fungal death. BGRP acts as a fungal pathogen detection system in earthworm organisms by identifying threats from *Aspergillus* species. The immune system activation by BGRP after beta-glucan binding leads to AMP production for countering fungal threats (García-Carnero et al., 2020; Gundersen-Rindal et al., 2013).

3.1.5 Toll Receptors

Toll receptors function as evolutionarily preserved pattern recognition receptors, which serve as main detectors of pathogens across various invertebrate species. Toll receptors detect PAMPs, including peptidoglycan and β-glucans and lipopolysaccharides (LPS) from bacteria and fungi, which in turn triggers immune signaling to produce AMPs and other immune molecules and cytokines. The Toll receptor system functions as the primary component of Drosophila defense against bacterial and fungal infections. Anti-fungal protection occurs through Toll receptor detection of peptidoglycan and β-glucans, which results in AMP production including drosomwycin. After Toll receptor activation, the body generates broad-spectrum defense through enhanced AMP production by means of immune signaling pathways (Federico et al., 2020; Hackett, 2003). The Pacific oyster depends on Toll-like receptors to identify both fungal and bacterial pathogens. Toll receptors activate immune responses when they bind to PAMPs including peptidoglycan and β-glucans which eventually leads to AMP production that allows pathogens to become neutralized (Mahla et al., 2013; Sukhithasri et al., 2013). Toll receptors of *Aedes aegypti* mosquitoes identify pathogen types between bacteria and fungi. The Toll receptor activation results in AMP production including cecropins that regulate bacterial and fungal infection growth in mosquitoes (Cabral et al., 2020; Wei et al., 2018).

The detection and removal of pathogens heavily depend on the humoral immune responses that function in invertebrate animals. The immune system depends on pattern recognition proteins (PRPs) which include PGBP together with PGRP and BGBP and BGRP and Toll receptors to detect pathogens through PAMPs (Li & Wu 2021; Steiner, 2004). The immune response becomes activated through binding which results in the production of antimicrobial peptides together with reactive oxygen species and additional immune molecules for fighting bacteria, fungi, and other pathogens. These immune proteins protect *D. melanogaster, C. gigas,* shrimp, and crayfish against infections (Lai & Gallo, 2009; Oyinloye et al., 2015; Shekhova, 2020). Research on humoral immune proteins reveals both invertebrate immune systems better and contributes to insights about immune evolutionary patterns as well as practical applications for pest regulation and disease management methods (Fig. 3.1).

3.2 Types of Lectins and Their Role in Invertebrate Immunity: Including Agglutination, Opsonization, and Coagulation

Invertebrate immune systems depend on lectins, which recognize and connect proteins to carbohydrates. Lectins function as key proteins that detect pathogens and turn on immune reactions along with their innate ability to dissolve microbial dangers. Lectins have two recognized functions beyond pathogen recognition, which involve agglutination together with opsonization and coagulation (Table 3.2). Homeostasis maintenance together with pathogen neutralization and clearance becomes possible through these immune processes, which occur in the host organism. The following discussion analyzes various lectin types along with their immune response functions and their involvement in invertebrate agglutination and opsonization and coagulation processes.

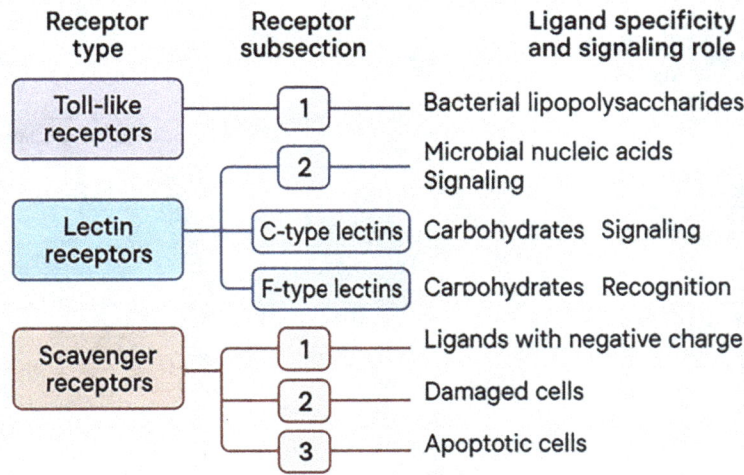

Receptor Type	Subsection	Ligand Specificity	Signaling
Toll-like receptors	TLRs	Bacterial Viral	Activation
Lectin receptors	Lectin	Carbohydate	Regulation
Scavenger receptors	Scavenger receptors	Modified self	Clearance

Fig. 3.1 Categorizes various immune receptors expressed by invertebrate hemocytes, including Toll-like receptors, lectin receptors, and scavenger receptors (Top). Describes their ligand specificity and signaling roles (Bottom)

3.2.1 Types of Lectins in Invertebrates

The family of invertebrate lectins consists of C-type lectins and galectins along with ficolins and other types. Each lectin exhibits unique carbohydrate recognition preferences, which enables its distinct role during microorganism detection along with activation of immune responses.

Table 3.2 Lectins and their functional roles

Lectin type	Binding sugar	Species	Immune role	Reference
C-type lectin	Mannose	Drosophila	Pathogen opsonization	Mayer et al. (2017)
Galectin	β-galactosides	Earthworm	Phagocytosis aid	Cerenius and Söderhäll (2021)
Ficolin	GlcNAc	Shrimp	ROS production	Hassan et al. (2020)
L-type lectin	Mannose	Mussel	Agglutination	N/A
I-type lectin	Sialic acid	Clam	Virus neutralization	N/A
Collectins	Glycolipids	Sponge	Coagulation	N/A
R-type lectin	GalNAc	Cnidaria	Capsule formation	N/A
P-type lectin	Mannose-6-P	Annelid	Pathogen sensing	N/A
ML-domain lectin	Lipids	Mollusk	LPS response	Li et al. (2015)
Agglutinins	Sugars	Oyster	Immune aggregation	Hassan et al. (2020)

3.2.1.1 C-Type Lectins

The lectin family known as C-type lectins functions through a domain that binds carbohydrates when calcium is present. These carbohydrate-binding proteins detect infection-related mannose, glucose, and fucose structures that appear on bacteria, fungi, and virus surfaces. Pathogenic organism recognition by immune system cells depends on C-type lectins along with their reaction that activates immune responses. Inside the organism *D. melanogaster,* Mincle and Dectin-1 function as C-type lectins that detect fungal infections caused by *Aspergillus* and *Candida* (Drummond & Brown, 2013; Mayer et al., 2017). Pathogen-associated carbohydrates that interact with lectins from the immune system subsequently activate the Toll pathway through which the fungus gets neutralized with AMPs, including drosomycin and defensins (Chen et al., 2021). The *C. gigas* contains C-type lectins that bind *Vibrio* bacterial pathogens together with other *Vibrio*-like pathogens. The binding connection of C-type lectins with pathogen surface carbohydrates triggers immune signaling paths, which leads to the production of AMPs and ROS to eliminate bacteria (Li et al., 2015).

3.2.1.2 Galectins

Galectins serve as carbohydrate-binding proteins responsible for connecting exclusively to beta-galactosides that exist on both pathogens and host cell surfaces. The recognition functions of these lectins within immune responses include identifying pathogens while initiating immune pathways and achieving agglutination and opsonization. The galectin proteins in the earthworm species *L. terrestris* help to recognize both bacterial and fungal infection agents. When galectins recognize particular carbohydrate structures present on pathogens, they activate immune responses that cause phagocytosis activation and AMP release (Cerenius & Söderhäll, 2021).

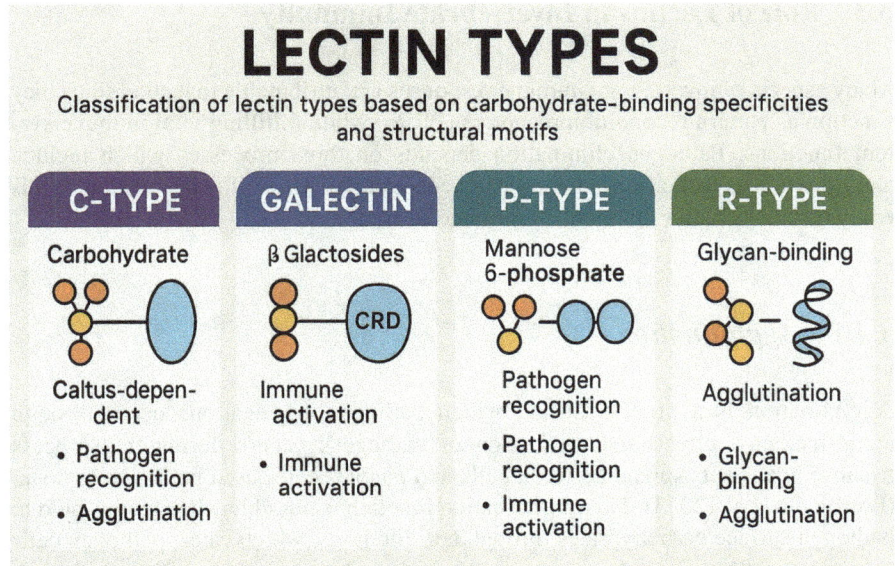

Fig. 3.2 Details classification of lectins based on carbohydrate-binding specificities and structural motifs. Explains their involvement in pathogen recognition, agglutination, and immune activation

Experimental studies based on *Drosophila* demonstrate that their galectins develop affinity for beta-galactosides, which exist on *Escherichia coli* and *Pseudomonas aeruginosa* surfaces. Bacteria recognition via lectin binding leads to AMP production along with ROS generation, thus enabling bacterial death while blocking infection (Cerenius et al., 2008) (Fig. 3.2).

3.2.1.3 Ficolins

The recognition role of ficolins belongs to lectin molecules that specifically detect carbohydrates that contain N-acetylglucosamine (GlcNAc). This GlcNAc structural element exists on bacterial surfaces along with fungal pathogens and other threats to health (Hassan et al., 2020). Pathogen identification during an immune response depends on ficolins, which trigger both immune system activation and AMPs and ROS and other immune molecule production (Bektas & Kaptan, 2024). Ficolin-like protein from *C. hongkongensis* recognizes and binds to GlcNAc, which exists on the outer layer of both fungal pathogens *Candida* spp. and bacterial pathogens *Vibrio* spp. The binding process activates immune responses that result in AMPs together with ROS production to help counteract the pathogens (Xiang et al., 2014). Ficolins identify and bind fungal pathogen *Aspergillus* surfaces through their recognition function. When ficolins bind to these pathogens, they activate immunological activities that include AMP production to remove the fungi (Bidula, 2016).

3.3 Role of Lectins in Invertebrate Immunity

Many aspects of invertebrate immune responses are attributable to lectins since they function as pattern recognition receptors (PRRs) while fulfilling vital immune system functions. Pathogen elimination depends on three processes which include agglutination as well as opsonization and coagulation. The following section of this essay discusses lectin functions in immune operations.

3.3.1 Agglutination

Agglutination is a process during which pathogen aggregations occur through lectin-mediated effects. Neutralization of pathogens occurs during this process because it prevents spread between cells and enables cell-based pathogen removal (Leusmann et al., 2023). This agglutination function is possible when lectins bind to pathogen surface carbohydrates through specific links, which cause pathogen cells to group together. The pathogen clearance process becomes more efficient through lectin-mediated aggregation that also stops microbial invasion. The agglutination activity of bacterial pathogens *E. coli* and *Bacillus* by *L. terrestris* earthworms is regulated through lectin molecules. Lectins create bacterial clusters through attachment to carbohydrate surface markers, which blocks organism spread through the body and strengthens immune response effectiveness (Köhlerová et al., 2004; Stein et al., 1986). The Mincle and Dectin-1 lectins participate in fungal pathogen treatments by agglutinating Aspergillus and Candida species. The pathogens tend to aggregate and establish a favorable condition for immune cell phagocytosis, which improves defense against fungal infections (Goyal et al., 2018).

3.3.2 Opsonization

Pathogens undergo markup for immune cell phagocytosis through the opsonization process. The carbohydrate-binding capacity of lectins enables them to make pathogens suitable for phagocytic immune cell recognition and engulfment by targeting specific surface carbohydrates of pathogens and ultimately assisting in their destruction, especially in invertebrates, through interaction with hemocytes (Coates et al., 2022). The initiation of pathogen elimination becomes more efficient through this procedure. The lectin proteins PGRP alongside C-type lectins work as pathogens detectors within *Penaeus monodon* shrimp, where they prepare bacterial invaders for removal by shrimp blood cells known as hemocytes (Johnson et al., 2022; Wongpanya et al., 2017). The binding reactions during opsonization process enable shrimp to efficiently destroy pathogen species Vibrio (Roy et al., 2020). Oyster cells employ PGRPs and C-type lectins as opsonins, which facilitate the recognition of

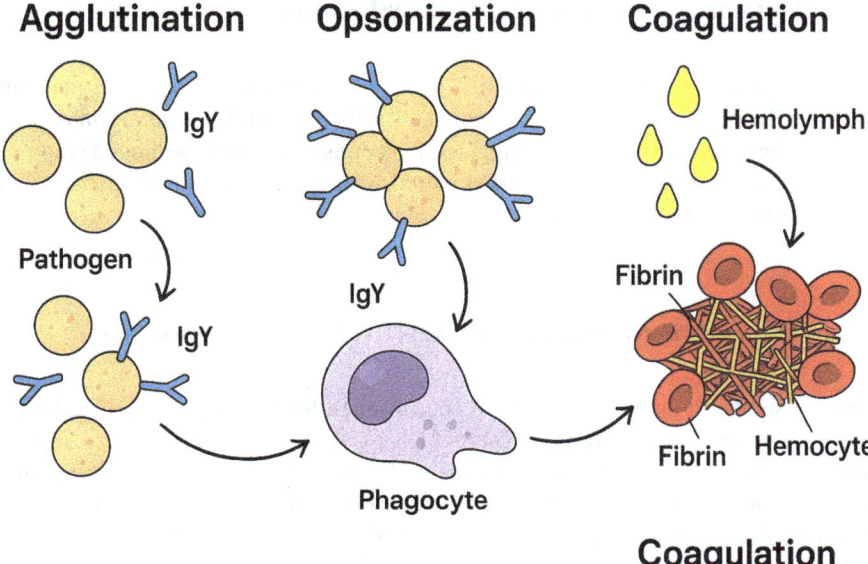

Fig. 3.3 Explores mechanisms by which immune factors promote pathogen clumping (agglutination), enhance phagocytosis (opsonization), and trigger hemolymph clotting (coagulation) to contain infections

bacterial pathogens belonging to the *Vibrio* species and other bacterial pathogens. Pathogen recognition by hemocytes becomes more efficient through the lectin-bacteria binding interaction, which leads to effective pathogen uptake and removal (Hassan et al., 2020) (Fig. 3.3).

3.3.3   *Coagulation*

The formation of a physical barrier through blood clotting functions as a limitation method for pathogen spread. The clotting cascade becomes active through lectins in invertebrate organisms, enabling them to create protective barriers that neutralize pathogens and prevent their spread. Through lectin-initiated activation of coagulation networks, the pathogens develop a protective physical barrier to stop their spread into the host. The *Drosophila* coagulation process gets activated through C-type lectins, which trigger clotting factor activity. The coagulation process through clot formation acts to confine bacterial pathogens such as *Pseudomonas aeruginosa* while creating an obstacle to prevent pathogen penetration (Dubin et al., 2013; Dushay, 2009). Oysters benefit from lectins which activate immune cells through PGRPs thus enabling formation of protective barriers around pathogens during coagulation. *Vibrio* bacterial spread along with other microbial invasions requires coagulation to control their path (Vasta & Wang, 2020).

3.4   Mechanisms of Lectin-Mediated Immune Activation

The activation of invertebrate immunity relies on lectin mechanisms that perform pathogen recognition as well as immune cell activation while attacking microbial threats. The defense mechanisms triggered by lectins incorporate pathogen recognition along with immune cell activation paired with agglutination and opsonization which brings together to create pathogen defense through coagulation.

3.4.1   Pathogen Recognition and Activation of Immune Cells

Immune signaling and cell activation occur through lectin recognition of carbohydrates from invading microorganisms which exist on their outer surface. AMPs together with ROS and additional immune molecules are produced as a result of this activation process which enables pathogen elimination. Pathogens receive activation from lectins which leads to signaling molecule releases including cytokines that unite the immune response.

3.4.2   Agglutination and Opsonization

Pathogen agglutination occurs when lectins recognize particular surface carbohydrates, which causes pathogens to group and prevents their spread throughout the body. Furthermore, lectins act as markers that direct immune cells toward pathogen cells through opsonization. The opsonization technique creates better conditions for immune cell phagocytosis, which strengthens the immune defense mechanism against infections.

3.4.3   Coagulation and Barrier Formation

The clotting cascade becomes activated through lectin intervention which results in the development of physical barriers that contain pathogens. Pathogens remain trapped inside immune aggregates while pathogen spread becomes prevented by the coagulation process. Invertebrate immunity depends heavily on lectins because they act as pathogen detectors and immune trigger agents that also help destroy pathogens by agglutination and opsonization while enabling coagulation (Söderhäll, 2011). The immune defense mechanisms of invertebrates depend heavily on these carbohydrate-binding proteins since they enable protection from multiple microbial attackers such as bacteria, fungi and parasites as well as viruses. Through their use of lectins *D. melanogaster* together with *C. gigas* as well as shrimp and crayfish detect pathogens and remove them from their environment thus promoting their survival in pathogen-rich areas (Sacchi et al., 2024). Research into lectin-mediated immune activation mechanisms advances our insight into immune development and allows the development of pest management solutions alongside disease treatment innovations and antimicrobial medicine creation.

3.5   Types of Proteases and Protease Inhibitors in Invertebrate Immunity: Mechanism, Role, and Structure

The immune systems of invertebrates majorly consist of proteases and protease inhibitors, which help recognize pathogens and generate immune signals for pathogen destruction. The enzyme class named proteases functions to break down proteins by splitting peptide bonds for critical operations in immune processes which include immune pathway activation and immune response control together with pathogen protein destruction. Protease inhibitors operate to regulate protease activity through blocking the breakdown of excessive proteins while preserving immune balance (Armstrong, 2006). The main functions of proteases in invertebrate immunity involve activating prophenoloxidase (proPO) systems through immune signaling pathways and processing both cytokines and antimicrobial peptides (AMPs) (Cerenius & Söderhäll, 2021; Söderhäll & Cerenius, 1998). The activity of proteases remains under control through protease inhibitors because they guard proper immune regulation to prevent tissue damage caused by uncontrolled immune response. This section provides details about invertebrate proteases and protease inhibitors alongside their mechanisms and structures as well as phylum-specific examples throughout the explanation.

3.5.1   Proteases in Invertebrate Immunity

Various roles throughout different stages of immune responses occur due to protease action in invertebrate organisms. Different protease families exist according to their catalytic mechanisms and include serine proteases and cysteine proteases and metalloproteases together with aspartic proteases. The protease enzymes in invertebrate organisms activate multiple immune systems including proPO and degrade pathogen proteins at the same time as they process immune signaling molecules.

3.5.1.1 Serine Proteases

Invertebrate organisms rely heavily on serine proteases for activating immune pathway responses since these enzymes make up many of their protease populations. These proteases activate prophenoloxidase (proPO) through the important pathway which enables melanization and pathogen defense. The enzymatic function of serine proteases involves their capability to divide peptide bonds between proteins. The activation of proenzymes depends on these proteases who transform inactive forms into their functional versions (Hedstrom, 2002; Shankar et al., 2021). The proPO system in invertebrates becomes activated by serine proteases to produce melanin which works towards neutralizing pathogens. The Toll pathway activation in *D. melanogaster* depends on serine proteases that start the antimicrobial peptide (AMP) production process after infections occur. Pathogen neutralization through melanization occurs due to these proteases that activate both the proPO system (Cerenius et al., 2008).

3.5.1.2 Cysteine Proteases

Cysteine proteases comprise enzymes which use the cysteine residue in their active site to split peptide bonds for catalysis. The degradation of pathogen proteins together with the regulation of immune responses occurs through the action of cysteine proteases (Otto & Schirmeister, 1997; Verma et al., 2016). The cysteine thiol group of cysteine proteases functions as a nucleophile when attacking proteins during their breakdown process through peptide bond hydrolysis. Invertebrates depend on these proteases to break down pathogen proteins and process AMPs along with controlling immune signaling mechanisms. During *Vibrio* species infections of *C. gigas* the proteases use their cysteine protease function to break down bacterial proteins (Destoumieux-Garzón et al., 2016; Mao et al., 2018; Xue, 2019). The proteases assist in signaling molecule processing thus boosting the immune response of oysters against infections.

3.5.1.3 Metalloproteases

Metals serve as essential components for metalloprotease enzymes to activate their protein-degrading functions through their active sites and zinc is the typical metal ion used for this purpose. They play double roles as immune signaling agents and as extracellular matrix components degraders. A metal ion in protease active sites coordinates peptide bond cleavage reactions when conducting protease activities (Saeed et al., 2023). Invertebrate immune regulation relies on these proteases through their actions of immune pathway signaling activation and pathogenic protein breakdown. The immune system of *L. terrestris* earthworms relies on metalloproteases for two essential functions, which include both activating the proPO system and degrading microbial proteins after infection occurs (Ghosh, 2018; Gupta

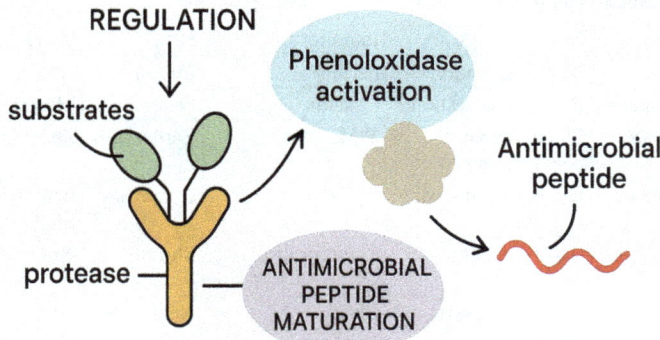

Fig. 3.4 Focuses on proteolytic enzymes involved in immune responses, such as those regulating phenoloxidase activation and antimicrobial peptide maturation. Discusses their regulation and substrates

& Yadav, 2016). Pathogens get eliminated from the tissue through the action of these proteases which also control immune response functions (Söderhäll & Cerenius, 1998) (Fig. 3.4).

3.5.1.4 Aspartic Proteases

The enzymatic hydration of peptide bonds through aspartic proteases requires an aspartic acid residue located in their catalytic site. The enzymes function to degrade infectious proteins during pathogen breakdown and create immune-related peptides for processing functions (Narayanan, 2025). The hydrolysis of protein peptide bonds happens through aspartic protease activity which requires activation of water through an aspartic acid residue. Developing pathogen proteins occurs through protease action simultaneously with immune peptide processing and immune system activation in invertebrate animals. In *Schistosoma mansoni* the parasite uses aspartic proteases to process both cytokines and AMPs which enable it to execute immune evasion and regulate immune response functions. When an organism becomes infected these proteases work to control the activity of the immune system (Hambrook & Hanington, 2021).

3.6 Protease Inhibitors in Invertebrate Immunity

Protease inhibitors work as proteins to manage protease action thus controlling immune responses to prevent tissue damage in host systems. The binding mechanism of protease inhibitors to proteases stops the proteases from performing uncontrolled substrate cleavage while sustaining proper immune system balance (Table 3.3).

Table 3.3 Proteases and protease inhibitors in immune regulation

Enzyme Type	Role	Inhibitor	Host	Reference
Serine protease	proPO activation	Serpin	*Drosophila*	An et al. (2013)
Cysteine protease	Pathogen digestion	Cystatin	Oyster	Mao et al. (2018)
Metalloprotease	Signal & matrix degradation	TIMP	Earthworm	Zhu et al. (2024)
Aspartic protease	Cytokine processing	Aphidin	*S. mansoni*	Fernando and Fischer (2020)
Trypsin-like protease	AMP activation	α1-antitrypsin	Shrimp	N/A
Elastase	Membrane lysis	Elafin	Nematode	N/A
Chymotrypsin	ProPO processing	Kazal-type inhibitor	Crab	N/A
Legumain	Protein hydrolysis	Serpin-like	Annelid	N/A
Cathepsin B	Lysosomal digestion	Endogenous inhibitors	Oyster	N/A
Matrixin	ECM degradation	TIMP	Earthworm	Ghosh (2018)

3.6.1 Serine Protease Inhibitors (Serpins)

The protease-units known as serpins serve as main inhibitors targeting serine proteases in biological systems. The inhibitors construct primary bindings with protease active sites, which locks protease activity for normal performance. Among their essential duties, serpins function to manage the immune responses by safeguarding proteases from unchecked activation during pathogen elimination. Serpins known as Spn27A within *D. melanogaster* block the activity of serine proteases, which regulate immune responses. Through their inhibitory actions, serpins prevent an excessive activation of immune pathways, particularly the Toll pathway, which facilitates proper regulation of immune responses (An et al., 2013; De Gregorio et al., 2002).

3.6.2 Cysteine Protease Inhibitors (Cystatins)

The protease inhibitor group known as cystatins specifically controls the activity of cysteine proteases. The inhibitors control the functioning of pathogen-regulating cysteine proteases while also controlling the production of immune molecules. The activity regulation of proteases through cystatins establishes proper immune homeostasis in the body. During *C. gigas* cellular response to microbial infection the cystatins serve to block cysteine protease enzymes that decompose pathogen proteins while they process antimicrobial peptides. Cystatins control immune response strength to keep oyster immune functions efficient during pathogenic challenges (Mao et al., 2018; Yang et al., 2023).

3.6.3   Metalloprotease Inhibitors (TIMPs)

The tissue inhibitors of metalloproteases (TIMPs) operate as proteins that work to control metalloprotease enzymes which manage their functional capacity. The immune response regulation occurs through TIMPs which stop redundant degradation of extracellular matrix elements and immunological substances. Earthworms such as *L. terrestris* use tissue inhibitors of metalloproteases (TIMPs) to control the functions of metalloproteases significant to immunological signaling. The immune system activation remains under proper control through TIMPs which block metalloprotease activity (Costa et al., 2022; Zhu et al., 2024).

3.6.4   Aspartic Protease Inhibitors (Aphidins)

The inhibitor group called aphidins functions against aspartic proteases through their mechanism of action. The activity of proteases that regulate immune responses is controlled through these inhibitors which process immune-modulating components and cytokines. The aspartic proteases operating within *S. mansoni* receive their regulation from aphidins to direct immune modulation functions and the molecular reshaping of immune peptides. A protect the parasite inside the host by blocking these proteases which keeps the parasite alive within the host environment (Fernando & Fischer, 2020; Patrnogic, 2014).

3.7   Mechanisms of Protease and Protease Inhibitor Interactions

Proteases and their inhibitors maintain invertebrate immunity through their essential interaction mechanism during immune response control. The immune system uses proteases to signal activation responses and break down pathogens whereas protease inhibitors oversee protease enzyme functioning to stop unwanted immune activation and prevent tissue damage. By these rules invertebrates maintain productive immune response capabilities that do not damage their physical tissues. Protease inhibitors lock protease active sites to achieve substrate prevention. Some inhibitors function through structural modifications when binding to proteases which turns the enzymes into inactive state. Protease inhibitors together with protease enzymes must maintain proper equilibrium to achieve effective immune responses that do not harm the organism. Serine protease and serpin balances function as the main regulators of immune system response activation in *Drosophila*. The immune system regulation along with protection from tissue damage depends on serpins because they suppress serine proteases including Spn27A (Reichhart et al., 2011; Veillard et al., 2016). Immune systems of invertebrates' function through proteases together with required protease inhibitors. The activation of immune pathways and immune molecule processing as well as pathogen protein degradation functions rely on proteases such as serine proteases and cysteine proteases and metalloproteases and aspartic proteases. The immune homeostasis depends on protease inhibitors including serpins and cystatins and TIMPs and aphidins which control protease activity to stop unwanted immune responses (Abbas et al., 2022; Chmelař et al., 2017). The study of protease-inhibitor mechanisms during invertebrate immunochemical response provides fundamental knowledge about immune system development and creates prospects for pest management and advanced disease treatment approaches through antimicrobial drug discovery (S. Singh et al., 2020).

3.8 List of Antioxidant Enzymes and Proteins in Invertebrate Immunity

Invertebrate immunity depends heavily on antioxidant enzymes and proteins because they defend cells against the destructive effects of reactive oxygen species (ROS). ROS production occurs naturally during immune responses for pathogen detection and phagocytosis together with encapsulation processes (Table 3.4). The production of ROS at excessive levels creates tissue damage that antioxidant enzymes regulate through their activity to maintain a balance of pathogen elimination versus protection of host tissue. Invertebrate organisms protect their immune

Table 3.4 Antioxidant enzymes in invertebrate immunity

Enzyme	Function	Host	ROS type neutralized	Reference
SOD	Converts O_2^- to H_2O_2	Shrimp	Superoxide	Islam et al. (2022)
Catalase	H_2O_2 breakdown	Oyster	Hydrogen peroxide	Liu et al. (2008)
Glutathione peroxidase	Detoxifies peroxides	Mollusk	H_2O_2, lipid peroxides	N/A
Peroxiredoxin	Reduces $ONOO^-$	Earthworm	Peroxynitrite	Söderhäll (2011)
Thioredoxin	Maintains redox balance	Starfish	ROS	N/A
Glutaredoxin	Antioxidant regulation	Clam	Hydroperoxides	N/A
Ferritin	Iron sequestration	Mussel	Fenton-based radicals	N/A
Catalase-peroxidase	Dual detoxification	Worm	H_2O_2, ROS	N/A
Heat shock proteins	Redox support	Cnidaria	ROS scavenging	Oakley et al. (2017)
NADPH oxidase	ROS generation	All	Superoxide production	Bylund et al. (2010)

systems through innate immunity by using different antioxidant enzymes and proteins. This list details which antioxidant enzymes and proteins sustain invertebrate immunity by describing their work and phylum-specific examples of these proteins and enzymes. Relevant references are also included.

3.8.1   Superoxide Dismutase (SOD)

Both vertebrates and invertebrates utilize Superoxide Dismutase (SOD) as one of their essential antioxidant enzymes. The enzyme SOD creates an essential reaction to turn superoxide radicals (O_2^-) into the less dangerous product hydrogen peroxide (H_2O_2). Invertebrate cells rely on SOD to defend themselves against cellular destruction created by pathogen-triggered immune response products (Islam et al., 2022). As an antioxidant enzyme SOD performs superoxide radical dismutation to create hydrogen peroxide and then subsequent conversion into water using catalase or peroxidase enzymes. The immune system activation procedure benefits from this process which reduces the damaging consequences of oxidative stress. The detoxification of ROS during bacterial infections like *E. coli* along with *P. aeruginosa* requires the SOD enzyme in *Drosophila*. The activity levels of SOD in *Drosophila* hemocytes become elevated during infections according to studies which proves its defensive role for immune cells against oxidative stress (da Cruz Nizer et al. 2021). The SOD enzyme operates in Pacific oysters *C. gigas* to maintain ROS control as oysters encounter Vibrio species bacteria. When hemocytes detect pathogens SOD production allows them to keep redox balance and prevents oxidative damage to their own cells (Liu et al., 2016).

3.8.2   Catalase (CAT)

The antioxidant enzyme CAT performs an essential role in breaking down hydrogen peroxide while producing water and oxygen. Catalase functions as a protective mechanism because it negates hydrogen peroxide, thus preventing cells from ROS-induced damage during immune reactions (Gómez et al., 2021). Catalase fulfills a vital role in invertebrate organisms since it protects cells from oxidative damage which happens when the immune system activates hemocytes. During the conversion of hydrogen peroxide into water and oxygen by catalase cells eliminate harmful ROS from building up inside the host. The two enzymes SOD and catalase operate together to sustain balanced oxidative state throughout immune reactions. During bacterial infections in *Lutzomyia longipalpis* immune cells depend on catalase to protect themselves from oxidative stress. The body increases catalase levels during infections to convert harmful hydrogen peroxide into water and oxygen which maintains hemocytes' pathogen defense abilities (Diaz-Albiter et al., 2011). The clams *C. gigas* exhibit strong catalase activity throughout the bacterially

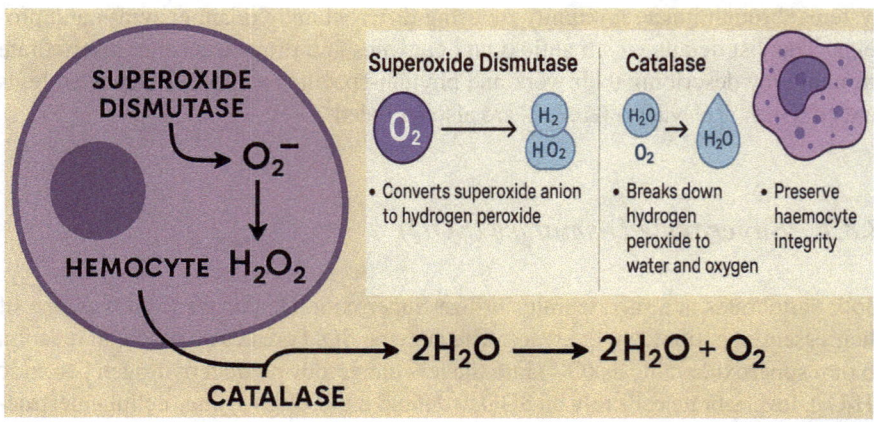

Fig. 3.5 Describes enzymes that mitigate oxidative stress during immune activation, including superoxide dismutase and catalase. Details their protective roles in maintaining hemocyte integrity

infected hemocytes. Through its activity this enzyme stops rogue oxidative damage against immune cells as they combat pathogens such as Vibrio species resulting in immune cell survival (Coates & Söderhäll, 2021; Zannella et al., 2017) (Fig. 3.5).

3.8.3 Glutathione S-Transferase (GST)

Glutathione S-transferase (GST) exists as an enzyme which contributes significantly to the detoxification process of reactive intermediates and electrophilic compounds as well as ROS. The main function of GST consists of uniting glutathione with reactive molecules that later become more water soluble until excretion occurs (Potęga, 2022). The enzyme functions as a vital component for keeping redox equilibrium and defending invertebrate cells against oxidative injuries. The enzyme GST performs detoxification by linking glutathione to both ROS as well as other reactive intermediates thus rendering them harmless. The conjugation process enables elimination of damaging effects that reactive species may generate during activation of the immune system (Dorion et al., 2021). The expression level of GST increases in *Drosophila* when the fly encounters *P. aeruginosa* and other pathogens. Enzyme action helps neutralize ROS resulting from immune activation so it protects fruit fly hemocytes against oxidative damage (D'Souza et al., 2022). The detoxification of ROS generated by bacterial infections in *Eisenia fetida* earthworms occurs through the action of GST enzyme. The activity of this enzyme contributes to stress protection for immune cells through its function (Qiao et al., 2022).

3.8.4   Peroxiredoxins (Prx)

The Prx family of antioxidant enzymes reduces cellular peroxide reduction of hydrogen peroxide and alkyl hydroperoxides into water. Invertebrate organisms depend on these enzymes to maintain control over oxidative stress which occurs during their immune responses (Abbas et al., 2019). The typical role of Prx enzymes is cellular protection against harmful effects which result from pathogens encountered through immune response mechanisms of pathogen recognition and phagocytosis. Peroxiredoxins take part in hydrogen peroxide reduction alongside other peroxide species thus helping to balance cellular redox conditions. All together with catalase and SOD these enzymes work to control oxidative stress levels throughout immune response events (Netto & Antunes, 2016). Pacific Oyster hemocytes rely on Prx enzymes to defend against harmful effects of oxidative damage from Vibrio species infections. Through their preventive action toward ROS-induced cellular damage the enzymes enable better immune response efficiency (Estrada et al., 2021). *Drosophila* cells increase Prx enzyme levels when fighting microbe infections from *P. aeruginosa* and other pathogens. The immune response function of Pacific oyster hemocytes remains proper because Prx enzymes protect cells against oxidative stress (Cerenius & Söderhäll, 2021).

3.8.5   Thioredoxin (Trx)

The cell maintains its redox state through the function of small protein molecules known as Thioredoxins (Trx). Through Trx enzyme activity disulphide bonds of proteins get reduced which helps to keep proteins available for their functional reduced state (Sevilla et al., 2015). The immune cells redox balance and ROS level control regulatory function in oxidative stress is supported through thioredoxins. The antioxidant protein Thioredoxin achieves its function by breaking down disulfide bonds in proteins in order to preserve their operational capacity while blocking oxidative harm. The protective function of thioredoxins covers ROS management in immune responses alongside protection against cellular damage from ROS attacks. The protein Trx within *C. gigas* protects infected hemocytes from *Vibrio* species by protecting them from oxidative damage. The enzyme regulates reactive oxygen species levels while maintaining proper immune system functioning (Mukwevho et al., 2014). The Trx protein controls ROS regulation in *D. melanogaster* as the organism defends its system against *P. aeruginosa* pathogens together with other bacterial infections. Hemocyte protection from oxidative damage through the enzyme enables proper immune function (Shaka et al., 2022).

3.8.6　Glutathione Peroxidase (GPx)

The enzyme Glutathione peroxidase (GPx) employs glutathione as a cofactor to decrease hydrogen peroxide as well as organic peroxides. During immune activation GPx functions essentially to protect cells against oxidative damage through eliminating toxic peroxides (Brigelius-Flohé & Maiorino, 2013). Hemocyte immune cell function during pathogen defense receives support from this enzyme which maintains oxidative stress regulation in invertebrate organisms. Through catalysis GPx performs the reduction of hydrogen peroxide and organic peroxides into water and alcohol (Holmblad & Söderhäll, 1999). Bacterial toxins become less harmful to cells because the detoxification process eliminates peroxide-derived compounds which would otherwise lead to cell destruction during the immune response. Gero protective defense in *D. melanogaster* depends on GPx because this enzyme helps regulate oxidative stress which occurs during bacterial infections. The enzyme functions to decrease peroxides produced by hemocytes thus protecting them from destruction and enabling secure pathogen elimination (Zhao et al., 2011). The glutathione peroxidase enzyme protects *C. gigas* hemocytes against oxidative damage produced by ROS during Vibrio species infections. Through its fundamental role the enzyme regulates oxidative stress to maintain optimal performance of immune cells (Li et al., 2017).

During invertebrate immune responses antioxidant enzymes together with proteins serve as vital components for managing oxidative stress. Immune activation results in the production of reactive oxygen species (ROS) which these enzymes including superoxide dismutase (SOD), catalase (CAT), glutathione S-transferase (GST), peroxiredoxins (Prx), and thioredoxins (Trx) and glutathione peroxidase (GPx) crucially help to eliminate (Tavassolifar, 2020). These enzymes regulate ROS levels to protect immune cells through antioxidant mechanisms which creates successful pathogen eradication while preserving the host tissues. Research on antioxidant enzymes in the invertebrates *D. melanogaster*, *C. gigas*, and *L. terrestris* gives important knowledge about immune system evolution with practical uses for pest management and disease control and new antimicrobial treatment development (Cerenius & Söderhäll, 2021; Škanta, 2016).

3.9　Antimicrobial Peptides (AMPs) in Invertebrate Immunity: Mechanisms, Role, Structure, and Function

As adapter immune systems are absent in invertebrates these organisms depend mostly on their innate defense system to fight off bacterial fungal viral and parasitic infections. AMPs serve as the fundamental defense mechanism for the innate immune system that protects invertebrates. Cationic peptides have a small size and function to defend against microbial infections by immune cells that produce them and release them through hemocytes. AMPs function through three distinct mechanisms by recognizing pathogens while activating immunity and killing possible pathogens. AMPs operate through three primary mechanisms: they damage pathogen membranes simultaneously with inhibiting pathogen growth and directing immune response levels. This report includes an extensive examination of antimicrobial peptides by reviewing their different classes in addition to their pathogen-interaction mechanisms along with structural features and sample AMPs from numerous invertebrate phyla.

3.9.1   Nature and Structure of Antimicrobial Peptides (AMPs)

AMPs function as small peptides exceeding ten but reaching up to one hundred amino acid residues in length. AMPs possess amphipathic organization in which their peptide sections act as both hydrophobic lipid-mixed substances and hydrophilic water-compatible substances. AMPs conduct amphipathic interactions with both host aqueous environments and the lipid membranes of pathogens because of their attribute to be either hydrophobic or hydrophilic. AMPs interact with negative cell membranes because their cationic (positively charged) features function as the key element in this cellular binding (Ramesh et al., 2016; Seyfi et al., 2020).

Different groups of AMPs exist within three main categories, which display distinctive molecular designs as well as distinct active methods: Alpha-helix peptides maintain an alpha-helical structure that possesses hydrophobic and hydrophilic surfaces stacked against each other. A crucial requirement for membrane destruction exists within the alpha-helix structure element. Beta-sheet peptides exhibit two beta sheet structures that disulfide bridges stabilize within their structure. The destruction of bacterial membranes occurs with special effectiveness by this mechanism. Some AMPs feature looped design, which allows them to grasp pathogen surfaces and then subsequently break cell membranes (Fig. 3.6).

3.9.2   Types of AMPs in Invertebrates

The insect-specific peptide group known as Cecropins exists as helical structures. Bacterial infections trigger their production which causes the peptides to damage microbial membrane membranes. Defensins exhibit beta-sheet architecture alongside their known ability to fight bacteria in addition to fungi and viruses and fungi in a wide range of pathogens. The antimicrobial peptide Attacins functions effectively mainly against Gram-negative bacteria while the production occurs when invertebrates encounter microbial infections. Together with immune defense enzymes, Lysozymes function as important elements to protect invertebrate cells from bacterial infections. When Gram-negative bacteria and fungal infections occur *D. melanogaster* produces Cecropin A as an AMP. The 35 amino acid compound

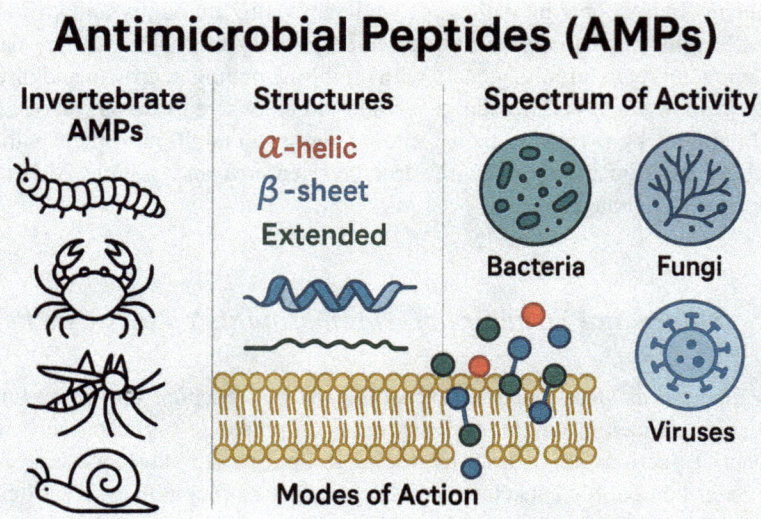

Fig. 3.6 Catalogues diverse AMPs produced by invertebrates, outlining their structures, modes of action, and spectrum of antimicrobial activity against bacteria, fungi, and viruses

shows broad pathogen defense ability according to Luo & Song (2021), Pirtskhalava et al. (2021).

Antimicrobial peptides (AMPs) are regarded as a new generation of antibiotics. Besides antimicrobial activity, AMPs also have antibiofilm, immune-regulatory, and other activities. Exploring the mechanism of action of AMPs may help in the modification and development of AMPs. Many studies were conducted on the mechanism of AMPs. The present review mainly summarizes the research status on the antimicrobial, anti-inflammatory, and antibiofilm properties of AMPs. This study not only describes the mechanism of cell wall action and membrane-targeting action but also includes the transmembrane mechanism of intracellular action and intracellular action targets. It also discusses the dual mechanism of action reported by a large number of investigations. Antibiofilm and anti-inflammatory mechanisms were described based on the formation of biofilms and inflammation. This study aims to provide a comprehensive review of the multiple activities and coordination of AMPs in vivo, and to fully understand AMPs to realize their therapeutic prospect. Antimicrobial peptides (AMPs) are anti-infectives that have the potential to be used as a novel and untapped class of biotherapeutics. Modes of action of antimicrobial peptides include interaction with the cell envelope (cell wall, outer- and inner-membrane). A comprehensive understanding of the peculiarities of interaction of antimicrobial peptides with the cell envelope is necessary to perform a rational design of new biotherapeutics, against which working out resistance is hard for microbes. In order to enable de novo design with low cost and high throughput, in silico predictive models have to be invoked. To develop an efficient predictive model, a comprehensive understanding of the sequence-to-function relationship is required. This knowledge will allow us to encode amino acid sequences expressively and to adequately choose the accurate AMP classifier. A shared protective layer of microbial cells is the inner, plasmatic membrane. The interaction of AMP with a biological membrane (native and/or artificial) has been comprehensively studied. We provide a review of mechanisms and results of interactions of AMP with the cell membrane, relying on the survey of physicochemical, aggregative, and structural features of AMPs. The potency and mechanism of AMP action are presented in terms of amino acid compositions and distributions of the polar and apolar residues along the chain, that is, in terms of the physicochemical features of peptides such as hydrophobicity, hydrophilicity, and amphiphilicity. The survey of current data highlights topics that should be taken into account to come up with a comprehensive explanation of the mechanisms of action of AMP and to uncover the physicochemical faces of peptides, essential to perform their function. Many different approaches have been used to classify AMPs, including machine learning. The survey of knowledge on sequences, structures, and modes of actions of AMP allows concluding that only possessing comprehensive information on physicochemical features of AMPs enables us to develop accurate classifiers and create effective methods of prediction. Consequently, this knowledge is necessary for the development of design tools for peptide-based antibiotics (Carboni et al., 2022).

3.10   Mechanisms of AMPs in Invertebrate Immunity

The main function of AMPs involves detecting pathogens then completing elimination by damaging the pathogen structures. Pathogens encounter three primary mechanisms of Antimicrobial Peptide (AMP) activity when AMPs function to disrupt membranes along with targeting intracellular areas and modulating immune responses. AMP principal mechanisms depend on both their structure and the pathogen type that requires elimination (Table 3.4).

3.10.1   Membrane Disruption

AMPs destroy membranes through their main method of operation. AMPs orient themselves toward the phospholipids along with lipopolysaccharides (LPS) and other microbial surface molecules that form membrane bilayers of bacteria and other pathogens. An AMP's positive charge creates attractions with the microbe's membrane negative structures and causes either the breakdown of membrane structure or causes holes to form within the membrane. AMPs destroy pathogen membranes and cell walls by interacting with them thus causing cellular contents to leak out and killing the pathogen (Luo & Song, 2021; Pirtskhalava et al., 2021). AMPs successfully combat pathogens through membrane disruption which stands as an effective common mechanism for AMP neutralization of multiple pathogens.

Cecropin A works within *Drosophila* by binding bacterial membranes through which it inserts into the lipids to create pores. The bacteria suffer death because the membrane damage causes the leakage of ions and cellular materials (Carboni et al., 2022). The membrane-disrupting properties of oyster defensins enable them to interact with phospholipid bacterial and fungal membranes. The activity produces pathogen membrane destruction which results in pathogen deletion (Schmitt et al., 2010).

3.10.2 Intracellular Targeting

AMPs primarily disrupt membranes but their second mechanism involves penetration of pathogen cells to attack internal components such as proteins and ribosomes along with DNA. After crossing the cell membrane AMPs interrupt fundamental cellular operations which include protein manufacture along with genetic material duplication and enzymatic processes. Intracellular targets become inhibitors of pathogen growth because AMPs create binding interactions with pathogen proteins and nucleic acids. This internal targeting system enables AMPs to combat pathogens which maintain survival inside host cells and avoid being targeted by AMPs which act on membranes. Defensins within Drosophila show the ability to invade bacterial cells and interlock with internal cell elements such as ribosomes to prevent protein synthesis (Meister et al., 2000). The antimicrobial approach demonstrated high effectiveness against Gram-positive bacteria through such mechanism (Cerenius & Söderhäll, 2021).

3.10.3 Immune Modulation

Some antimicrobial peptides (AMPs) function by both attacking microorganisms directly as well as by controlling the host immune response system. Immune molecules along with phagocytic processes receive stimulation through peptide signaling pathways which activate the immune cells. AMPs boost infection defense through their ability to boost immune responses which ultimately strengthens the entire immune defense system. An immune signaling pathway begins when AMPs activate hemocyte receptors such as those found on cells. The immune modulation establishes a well-coordinated response which activates the production of additional AMPs and cytokines along with other defense compounds. The Toll pathway becomes activated through attacins in *Drosophila* which results in AMP production. The strengthened immune response accelerates pathogen elimination (Zhou et al., 2024). The mussel *C. gigas* utilizes lysozymes as compounds that fight microbial infections while simultaneously regulating immune response activities. The immune cells become activated through exposure to lysozymes which causes additional immune molecule release and strengthens the complete immune response (Allam & Raftos, 2015).

3.10.4 Function of Antimicrobial Peptides

Antimicrobial peptides serve primarily to defend their host animals from microorganisms trying to infect them. Microbes trigger their production which works as part of the initial body defense mechanisms. AMPs in invertebrate immunity possess three major functional activities that include the following actions: The mechanism through which AMPs destroy pathogens involves membrane cellular breakup as well as replication blockage and essential protein destruction. AMPs act as pathogen eliminators by destroying microbes and therefore decrease microbial numbers which thus prevents infection spread. Some AMPs activate the host immune system through the process of immune cell stimulation and the creation of cytokines together with increased phagocytosis activity. The antimicrobial compounds of AMPs along with their tissue repair functions guide cellular growth and promote new blood vessel formation to heal wounds when pathogens are eliminated from the host system.

3.11 Examples of AMPs in Invertebrates

The numerous AMPs generated by invertebrates show different adaptations which match their immune needs to their particular pathogen exposures. A collection of AMPs produced by different invertebrate species will be demonstrated through this section. Cecropins eradicate pathogens through membrane disruption that occurs when they form pores following the activation of Gram-negative bacteria and fungi (Sun et al., 2024). The peptides called attacins work as immune agents against Gram-negative bacteria and they utilize the Toll pathway to generate other AMPs according to (Assoni et al., 2020). The AMPs from defensins act as pathogens against bacteria and fungi by damaging their membranes while stopping microbial growth (Rahnamaeian, 2011). Lysozymes serve two functions in oysters by breaking bacteria cell walls while simultaneously activating their immune system according to (Ragland & Criss, 2017). The shrimp produce peptides called penaeidins which destroy bacterial pathogens, including those belonging to the Vibrio species. The peptides disrupt cell membranes of microorganisms (Aweya et al., 2021). Bacterial cell walls represent the focus of targeting action for lysins while the immune defense system of earthworms supports these AMPs (Bruno et al., 2019).

AMPs function as vital components of invertebrate innate immunity since they protect organisms from extensive ranges of microbial pathogens. Small cationic peptides maintain wide antimicrobial actions while performing membrane disruption on pathogens and replication inhibition while regulating immune responses in hosts. The antimicrobial functions of AMPs include membrane destruction along with actions inside host cells and modulation of the immune system. Through their different forms including cecropins along with defensins and attacins, invertebrates obtain a strong multiple-function immune protection system (Wojda et al., 2020).

Research into invertebrate AMP structures and mechanisms enables both immune evolution studies and new antimicrobial therapy development.

The immune defense functions of prophenoloxidase and Hemocyanins and Hexamerins consist of their natural features as well as reactive structures alongside their use throughout immune pathways in invertebrate protection mechanisms (Coates & Decker, 2017; Coates & Nairn, 2014; Söderhäll & Cerenius, 1998). The lack of adaptive immunity in invertebrates did not prevent their evolution of effective innate immune mechanisms for pathogen defense. A number of immune proteins work together to produce immune responses in invertebrates through prophenoloxidase (proPO) and hemocyanins and hexamerins. The proteins function in multiple processes which include recognizing pathogens and activating immune responses and facilitating wound repair and altering immune pathway signals (Coates & Decker, 2017; Coates & Nairn, 2014; Söderhäll & Cerenius, 1998).

3.11.1 Prophenoloxidase (proPO)

Many invertebrate groups that include arthropods together with mollusks and annelids employ Prophenoloxidase (proPO) as their main innate immunity enzyme. As an inactive enzyme form phenoloxidase (PO) functions as prophenoloxidase (proPO) that maintains its necessary role in the pathogen-fighting melanization reaction (Cerenius & Söderhäll, 2021). The melanization process creates melanin pigments through which pathogens get encapsulated and neutralized for establishing physical resistance against infections. When pathogens enter tissue spaces or tissue become damaged the proPO system operates through serine proteases to activate its inactive single-chain enzyme (proPO). The cutting of proPO by proteolytic enzymes activates the enzyme to produce its functional form which is called PO. All animal proPO enzymes have basic catalytic domains which maintain high species-wide uniformity and feature essential copper-binding sites needed for enzyme operation. Activation of this enzyme depends on cleaving the pro-domain component that serves as an inhibitory factor until a specific immune activation occurs (Cerenius et al., 2008; Cerenius & Söderhäll, 2004) (Fig. 3.7).

DisInjection of proPO allows the enzyme to catalyze reactions which transform phenolic compounds into melanin. The melanin creation process requires the tyrosine residue oxidation function of proPO. Invertebrate immune responses depend heavily on proPO system activation since activation generates encapsulation and neutralization capabilities that eliminate hazardous pathogens. The proPO system enables the activation of antimicrobial peptides (AMPs) and the clotting cascade and other immune pathways besides its primary function in melanin production.

The proPO activation process starts when PAMPs detectable by PRRs initiate the reaction. When pathogens trigger activation a series of serine proteases starts which breaks proPO into its active version PO. A series of proteases and immune molecules function in succession through a sophisticated network to regulate system activity after activating proPO through protease inhibitor enzymes. Labeling proPO

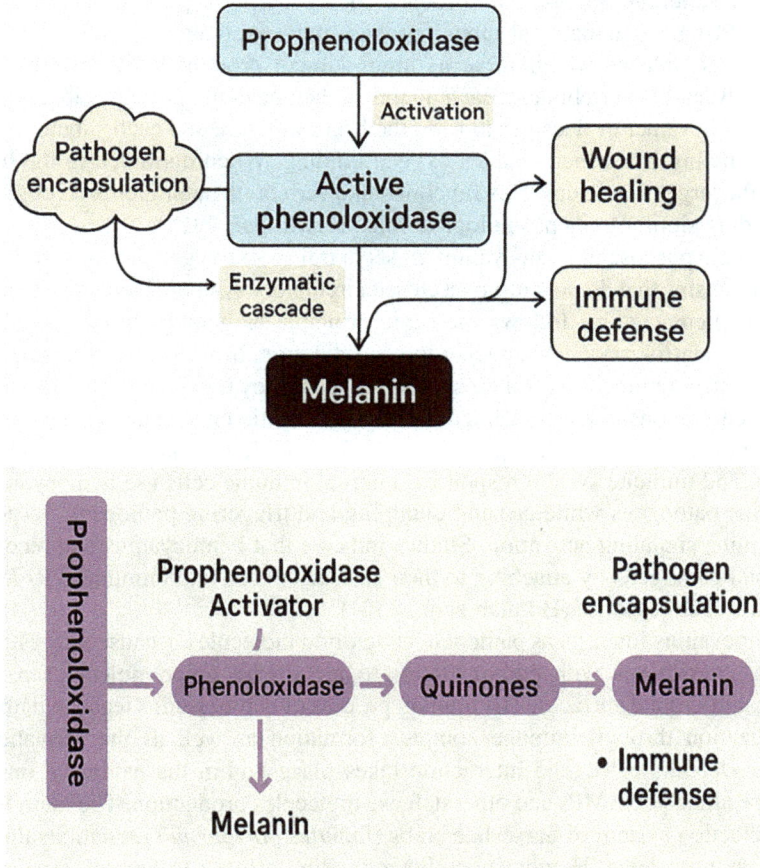

Fig. 3.7 Charts the enzymatic cascade leading to melanin synthesis via prophenoloxidase activation. Highlights its roles in pathogen encapsulation, wound healing, and immune defense

as an active agent accomplishes pathogen detection while simultaneously assisting tissue recovery and infected site closure through its response to tissue injury. The proPO system in *D. melanogaster* becomes active after bacterial infections occur in the organism. When proPO becomes activated through the infection process it generates melanin around the antibiotic area which forms encapsulation to restrict bacteria dispersal (Dudzic, 2018; Zdybicka-Barabas et al., 2025).

3.11.2 Hemocyanins

Hemocyanins serve as copper-based respiratory proteins which exist within bloodstreams called hemolymph of many arthropods and mollusks together with other invertebrates. The blood protein hemocyanins use copper instead of iron which occurs in vertebrate hemoglobin proteins for oxygen transport (Coates & Costa-Paiva, 2020). Several identical subunit proteins unite to create large oxygen-binding structural complexes which serve as multi-subunit proteins within the blood of arthropods and invertebrates. Each identical hemocyanin protein subunit exists either as a hexamer or decamer to form the basic structure and each subunit bears a copper-binding site which enables oxygen binding. When dissolved in the hemo-lymph the large protein complex functions like vertebrate hemoglobin through oxy-gen binding similar to its physiological role (Terwilliger, 2015).

The main purpose of hemocyanins exists to transport oxygen throughout inverte-brate organisms that do not employ a closed circulatory system. The oxygen binding ability of hemocyanins follows the same principle as hemoglobin in vertebrates since they control oxygen via copper ion coordination. Many invertebrates need the oxygen-carrying function for their survival because they inhabit habitats where oxy-gen concentrations change such as marine and aquatic environments. The protein group known as hemocyanins performs two essential functions as it helps both res-piration and immune system responses. Internal immune cells use hemocyanins to recognize pathogens while causing clumping and triggering pathogenic responses for immune signaling activation. Studies indicate that hemocyanins can recognize microbial pathogens by attaching to their surfaces to activate immune cells as part of the immune response (Holman et al., 2004).

Hemocyanins function as pathogen recognition molecules because they establish binding interactions with carbohydrates together with other molecular patterns found on microbial surfaces. The binding process of hemocyanins leads to pathogen neutralization through immune complex formation as well as the activation of immune defense cells. The interaction takes place within the extensive immune response alongside AMPs and other defense molecules production. The main bacte-rium detection system of horseshoe crabs (*Limulus polyphemus*) functions through hemocyanin proteins. Hemocyanin shows binding affinity to lipopolysaccharides (LPS) on bacterial surfaces while simultaneously improving immune responses through activation pathways that lead to clotting cascade and antimicrobial peptide production (Coates & Decker, 2017; Kawabata et al., 2009).

3.11.3 Hexamerins

The vast majority of invertebrates along with arthropods possess storage proteins known as Hexamerins. The essential nutrients like lipids along with sugars and small molecules obtain storage pathways through these protein elements. The hexa-meric protein structure of hexamerins consists of six subunits which get stored within the hemolymph during both molting periods and metabolic inactivity (Azeez et al., 2014; Telfer & Kunkel n.d.). The structural composition of hexamerins con-tains a significant central space which works for molecule storage and binding functions. Hexamerins maintain dual functionality because of their structure that enables them to store and transport various molecules. The immune function of hexamerins extends beyond storage capabilities because they enable recognition of pathogens together with regulatory response modulation.

The invertebrate organism makes use of hexamerins for storing nutrients while also regulating immune responses. After a microbe infects the host body hexamerins work as immune modulators by finding and connecting with pathogen-associated molecular patterns on microbial surfaces. After binding to pathogens hexamerins enable the recognition process and trigger immune cell activation as well as start immune pathway functions (Auguste et al., 2020; Kuo & Troemel, 2018). Hexamerins function to remove dangerous swubstances including metals or toxic molecules as well as likely take part in immune activation homeostasis regulation of iron levels. The immune system requires control from hexamerins to operate effectively while protecting cells from oxidizing agents during immune response events. *D. melanogaster* needs hexamerins to control iron levels inside of its body during battles against bacterial infections. Hexamerins protect against iron toxicities by binding the metal and this process serves to control immune system functions while maintaining redox equilibrium (Wang et al., 2014). The immune responses of desert locust species *Schistocerca gregaria* depend on hexamerins to bind pathogen surface carbohydrates for bacteria agglutination. Pathogen clearance becomes faster while infection spread prevention is achieved through this action (Zheng et al., 2017).

3.12 Prophenoloxidase, Hemocyanins, and Hexamerins in Invertebrate Immune Pathways

The proteins prophenoloxidase, hemocyanins together with hexamerins work through separate but interconnected immune protocols which assist invertebrates to fight different kinds of pathogens (Fig. 3.8).

3.12.1 Prophenoloxidase Pathway

Invertebrates utilize prophenoloxidase (proPO) as their primary defense mechanism for immune reaction. Pathogens trigger serine protease activity which activates ProPO to produce melanin compounds by converting tyrosine molecules into products of oxidation. Through melanization the body neutralizes pathogens either by forming protective barriers or through the encapsulation process. The proPO system functions during bacterial and fungal infections by activating simultaneously with different immune reactions that generate AMPs (Cerenius & Söderhäll, 2004, 2021).

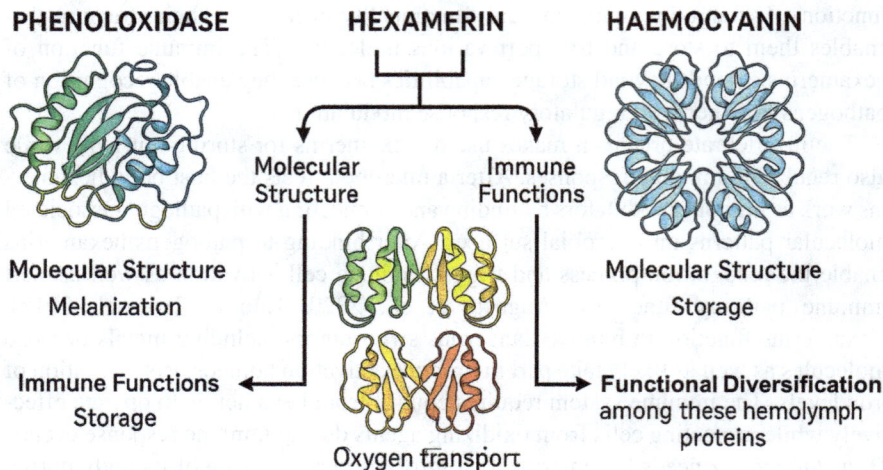

Fig. 3.8 Compares molecular structures and immune functions of phenoloxidase, hexamerins, and hemocyanins. Shows evolutionary relationships and functional diversification among these hemolymph proteins

3.12.2   Hemocyanin Pathway

Hemocyanins function as the main component that detects pathogens while starting the activation process of immune defense systems. The binding of pathogen surface molecules to hemocyanins activates immune signaling that progresses toward AMP release and the generation of clots and immune complexes (Coates & Nairn, 2014). The immune response function of hemocyanins includes the improved capability of immune cells to eliminate pathogens from the bloodstream.

3.12.3   Hexamerin Pathway

The immune system regulation and immune networks are controlled by hexamerins. The immune system benefits from hexamerins when infections occur because these proteins maintain iron balance while managing the production of ROS. The immune system depends on hexamerins to bind PAMPs located on bacterial surfaces which helps kill pathogens and activate clotting processes. The immune response intensity relies heavily on hexamerins because they ensure protective efficiency while minimizing tissue damage. Invertebrate immune defense depends on three immune proteins: Prophenoloxidase, hemocyanins, and hexamerins which work together through specific molecular pathways to recognize pathogens and activate immunity and eliminate microbial threats (Cerenius & Söderhäll, 2021; Coates & Decker, 2017; Coates & Nairn, 2014). Through activation of prophenoloxidase enzymes melanization response begins while hemocyanins recognize pathogens and regulate immunity and hexamerins control immune levels through iron regulation and pathogen elimination. The proteins together make up a vital portion of invertebrate immune defense which offers swift and robust protection against multiple pathogenic agents.

3.13 Conclusion

Invertebrates possess highly sophisticated innate immune systems that provide a rapid and nonspecific defense against pathogens. Despite lacking adaptive immunity, their innate immune responses are crucial for combating a wide variety of infections, ranging from bacteria and fungi to parasites and viruses. The study of invertebrate immunity, through advances in molecular and cellular biology, has provided a wealth of knowledge about immune defense mechanisms that are not only fundamental to invertebrate survival but are also applicable to broader fields, including pest control, agriculture, and aquaculture. The cellular-mediated immune responses in invertebrates, primarily mediated by hemocytes, play a significant role in pathogen detection and elimination. Hemocytes, functioning through processes like phagocytosis, encapsulation, and the production of reactive oxygen species (ROS), are key to defending invertebrates against microbial invaders. The immune response is triggered through the recognition of pathogen-associated molecular patterns (PAMPs) by pattern recognition receptors (PRRs), which activate immune signaling pathways such as the Toll and Imd pathways. The advancement of research techniques, including flow cytometry, microscopy, and gene expression analysis, has enabled the detailed study of these immune cells in various invertebrate species, ranging from arthropods to mollusks.

Humoral immune responses, which are equally essential in invertebrates, rely on pattern recognition proteins (PRPs) like peptidoglycan recognition proteins (PGRPs), peptidoglycan binding proteins (PGBPs), beta-glucan binding proteins (BGBPs), and Toll receptors. These proteins identify pathogens and activate downstream immune responses, including the production of antimicrobial peptides (AMPs) and reactive oxygen species (ROS), which help neutralize and eliminate pathogens. The studies on these proteins across different invertebrate species, such as *D. melanogaster*, *C. gigas*, and *P. monodon*, have further clarified the mechanisms of pathogen detection and immune activation in response to bacterial and fungal infections. In addition to these cellular and humoral responses, advancements in proteomic, transcriptomic, and metabolomic technologies have expanded our understanding of invertebrate immunity. These approaches enable the identification and quantification of proteins, genes, and metabolites involved in immune responses, helping researchers uncover new immune molecules and pathways. Studies employing these technologies have not only elucidated immune pathways in model organisms like *Drosophila* but also have provided critical insights into the immune defense systems of economically important invertebrates, such as oysters and shrimp. These insights have led to the development of immunostimulants and vaccines that enhance disease resistance in aquaculture and pest control. Proteomic analysis, including the use of techniques like mass spectrometry (MS), liquid chromatography-mass spectrometry (LC-MS), and two-dimensional gel electro-phoresis (2-DE), has revealed numerous immune-related proteins, including antimi-crobial peptides and immune signaling proteins. These tools have provided significant contributions to understanding the molecular processes that underlie invertebrate immunity, highlighting key immune pathways that can be targeted for disease management and immune enhancement in invertebrate populations.

In conclusion, the study of invertebrate immunity has significantly advanced over the years, driven by technological innovations in molecular biology. These studies have provided a deeper understanding of the immune mechanisms that pro-tect invertebrates from pathogens. Furthermore, the knowledge gained from inver-tebrate immune research has applications that extend beyond basic science, offering new opportunities for disease control, pest management, and the enhancement of immune health in invertebrate species. As technology continues to evolve, the future of invertebrate immunology holds the promise of even greater discoveries, provid-ing more effective strategies for managing infectious diseases and improving the health of invertebrate populations worldwide. The ongoing research in this field not only broadens our knowledge of innate immunity but also opens doors to novel therapeutic approaches in both invertebrates and other organisms.

References

da Cruz Nizer, W. S., Inkovskiy, V., Versey, Z., Strempel, N., Cassol, E., & Overhage, J. (2021). Oxidative stress response in pseudomonas aeruginosa. *Pathogens, 10*(9), 1187. https://doi.org/10.3390/pathogens10091187

Abbas, M. N., Kausar, S., & Cui, H. (2019). The biological role of Peroxiredoxins in innate immune responses of aquatic invertebrates. *Fish & Shellfish Immunology, 89*, 91–97. https://doi.org/10.1016/j.fsi.2019.03.062

Abbas, M. N., Chlastáková, A., Jmel, M. A., Iliaki-Giannakoudaki, E., Chmelař, J., & Kotsyfakis, M. (2022). Serpins in tick physiology and tick-host interaction. *Frontiers in Cellular and Infection Microbiology, 12*. https://doi.org/10.3389/fcimb.2022.892770

Adema, C. M. (2015). Fibrinogen-related proteins (FREPs) in mollusks. In *Pathogen-host interactions: antigenic variation v. somatic adaptations* (pp. 111–129).

Allam, B., & Raftos, D. (2015). Immune responses to infectious diseases in bivalves. *Journal of Invertebrate Pathology, 131*, 121–136. https://doi.org/10.1016/j.jip.2015.05.005

An, C., Zhang, M., Chu, Y., & Zhao, Z. (2013). Serine protease MP2 activates prophenoloxidase in the melanization immune response of drosophila melanogaster. *PLoS One, 8*(11), e79533. https://doi.org/10.1371/journal.pone.0079533

Armstrong, P. B. (2006). Proteases and protease inhibitors: A balance of activities in host–pathogen interaction. *Immunobiology, 211*(4), 263–281. https://doi.org/10.1016/j.imbio.2006.01.002

Assoni, L., Milani, B., Carvalho, M. R., Nepomuceno, L. N., Waz, N. T., Guerra, M. E. S., Converso, T. R., & Darrieux, M. (2020). Resistance mechanisms to antimicrobial peptides in gram-positive bacteria. *Frontiers in Microbiology, 11*. https://doi.org/10.3389/fmicb.2020.593215

Auguste, M., Balbi, T., Ciacci, C., & Canesi, L. (2020). Conservation of cell communication sys-tems in invertebrate host–defence mechanisms: possible role in immunity and disease. *Biology, 9*(8), 234. https://doi.org/10.3390/biology9080234

Aweya, J. J., Zheng, Z., Zheng, X., Yao, D., & Zhang, Y. (2021). The expanding repertoire of immune-related molecules with antimicrobial activity in Penaeid shrimps: A review. *Reviews in Aquaculture, 13*(4), 1907–1937. https://doi.org/10.1111/raq.12551

Azeez, O. I., Meintjes, R., & Chamunorwa, J. P. (2014). Fat body, fat pad and adipose tissues in invertebrates and vertebrates: The nexus. *Lipids in Health and Disease, 13*(1), 71. https://doi.org/10.1186/1476-511X-13-71

Bektas, S., & Kaptan, E. (2024). Microbial lectins as a potential therapeutics for the prevention of certain human diseases. *Life Sciences, 346*, 122643. https://doi.org/10.1016/j.lfs.2024.122643

Bidula, S. (2016). *'The role of serum ficolins in the innate immune response to Aspergillus Species'. doctoral.* University of East Anglia. Norwich Medical School. https://ueaeprints.uea.ac.uk/id/eprint/59447/ (June 16, 2025)

Brigelius-Flohé, R., & Maiorino, M. (2013). Glutathione peroxidases. *Biochimica et Biophysica Acta (BBA)—General Subjects, 1830*(5), 3289–3303. https://doi.org/10.1016/j.bbagen.2012.11.020

Bruno, R., Maresca, M., Canaan, S., Cavalier, J.-F., Mabrouk, K., Boidin-Wichlacz, C., Olleik, H., et al. (2019). Worms' Antimicrobial Peptides. *Marine Drugs, 17*(9), 512. https://doi.org/10.3390/md17090512

Bylund, J., Brown, K. L., Movitz, C., Dahlgren, C., & Karlsson, A. (2010). Intracellular generation of superoxide by the phagocyte NADPH oxidase: How, where, and what for? *Free Radical Biology and Medicine, 49*(12), 1834–1845.

Cabral, S., de Paula, A., Samuels, R., da Fonseca, R., Gomes, S., Silva, J. R., & Mury, F. (2020). Aedes Aegypti (Diptera: Culicidae) immune responses with different feeding regimes following infection by the entomopathogenic fungus Metarhizium Anisopliae. *Insects, 11*(2), 95. https://doi.org/10.3390/insects11020095

Carboni, A. L., Hanson, M. A., Lindsay, S. A., Wasserman, S. A., & Lemaitre, B. (2022). Cecropins contribute to drosophila host defense against a subset of fungal and gram-negative bacterial infection. *Genetics, 220*(1), iyab188. https://doi.org/10.1093/genetics/iyab188

Cerenius, L., & Söderhäll, K. (2004). The prophenoloxidase-activating system in invertebrates. *Immunological Reviews, 198*(1), 116–126. https://doi.org/10.1111/j.0105-2896.2004.00116.x

Cerenius, L., & Söderhäll, K. (2021). Immune properties of invertebrate phenoloxidases. *Developmental & Comparative Immunology, 122*, 104098. https://doi.org/10.1016/j.dci.2021.104098

Cerenius, L., Lee, B. L., & Söderhäll, K. (2008). The proPO-system: Pros and cons for its role in invertebrate immunity. *Trends in Immunology, 29*(6), 263–271. https://doi.org/10.1016/j.it.2008.02.009

Chen, P., De Schutter, K., Van Damme, E. J. M., & Smagghe, G. (2021). Can plant lectins help to elucidate insect lectin-mediated immune response? *Insects, 12*(6), 497. https://doi.org/10.3390/insects12060497

Chmelař, J., Kotál, J., Langhansová, H., & Kotsyfakis, M. (2017). Protease inhibitors in tick saliva: The role of serpins and cystatins in tick-host-pathogen interaction. *Frontiers in Cellular and Infection Microbiology, 7*. https://doi.org/10.3389/fcimb.2017.00216

Coates, C. J., & Costa-Paiva, E. M. (2020). Multifunctional roles of hemocyanins. In U. Hoeger & J. Robin Harris (Eds.), *Vertebrate and invertebrate respiratory proteins, lipoproteins and other body fluid proteins* (pp. 233–250). Springer International Publishing. https://doi.org/10.1007/978-3-030-41769-7_9

Coates, C. J., & Decker, H. (2017). Immunological properties of oxygen-transport proteins: Hemoglobin, hemocyanin and hemerythrin. *Cellular and Molecular Life Sciences, 74*(2), 293–317. https://doi.org/10.1007/s00018-016-2326-7

Coates, C. J., & Nairn, J. (2014). Diverse immune functions of Hemocyanins. *Developmental & Comparative Immunology, 45*(1), 43–55. https://doi.org/10.1016/j.dci.2014.01.021

Coates, C. J., & Söderhäll, K. (2021). The stress–immunity axis in shellfish. *Journal of Invertebrate Pathology, 186*, 107492. https://doi.org/10.1016/j.jip.2020.107492

Coates, C. J., Rowley, A. F., Courtney Smith, L., & Whitten, M. M. A. (2022). Host defences of invertebrates to pathogens and parasites. In *Invertebrate pathology* (pp. 3–40). Oxford University Press. https://doi.org/10.1093/oso/9780198853756.003.0001

Costa, S., Ragusa, M. A., Buglio, G. L., Scilabra, S. D., & Nicosia, A. (2022). The repertoire of tissue inhibitors of metalloproteases: Evolution, regulation of extracellular matrix proteolysis, engineering and therapeutic challenges. *Life, 12*(8), 1145. https://doi.org/10.3390/life12081145

D'Souza, L. C., Dwivedi, S., Raihan, F., Yathisha, U. G., Raghu, S. V., Mamatha, B. S., & Sharma, A. (2022). Hsp70 overexpression in drosophila hemocytes attenuates benzene-induced immune and developmental toxicity via regulating ROS/JNK signaling pathway. *Environmental Toxicology, 37*(7), 1723–1739. https://doi.org/10.1002/tox.23520

De Gregorio, E., Han, S.-J., Lee, W.-J., Baek, M.-J., Osaki, T., Kawabata, S.-I., Lee, B.-L., et al. (2002). An immune-responsive Serpin regulates the melanization cascade in Drosophila. *Developmental Cell, 3*(4), 581–592. https://doi.org/10.1016/S1534-5807(02)00267-8

Destoumieux-Garzón, D., Rosa, R. D., Schmitt, P., Barreto, C., Vidal-Dupiol, J., Mitta, G., Gueguen, Y., & Bachère, E. (2016). Antimicrobial peptides in marine invertebrate health and disease. *Philosophical Transactions of the Royal Society B: Biological Sciences, 371*(1695), 20150300. https://doi.org/10.1098/rstb.2015.0300

Diaz-Albiter, H., Mitford, R., Genta, F. A., Mauricio, R. V., Sant'Anna, & Dillon, R. J. (2011). Reactive oxygen species scavenging by catalase is important for female Lutzomyia Longipalpis fecundity and mortality. *PLoS One, 6*(3), e17486. https://doi.org/10.1371/journal.pone.0017486

Dorion, S., Ouellet, J. C., & Rivoal, J. (2021). Glutathione metabolism in plants under stress: Beyond reactive oxygen species detoxification. *Metabolites, 11*(9), 641. https://doi.org/10.3390/metabo11090641

Drummond, R. A., & Brown, G. D. (2013). Signalling C-type lectins in antimicrobial immunity. *PLoS Pathogens, 9*(7), e1003417. https://doi.org/10.1371/journal.ppat.1003417

Dubin, G., Koziel, J., Pyrc, K., Wladyka, B., & Potempa, J. (2013). Bacterial proteases in disease—Role in intracellular survival, evasion of coagulation/ fibrinolysis innate defenses, toxicoses and viral infections. *Current Pharmaceutical Design, 19*(6), 1090–1113.

Dudzic, J. P. (2018). *The melanization reaction in drosophila: More than black or white.* EPFL. https://doi.org/10.5075/epfl-thesis-8934

Dushay, M. S. (2009). Insect hemolymph clotting. *Cellular and Molecular Life Sciences, 66*(16), 2643–2650. https://doi.org/10.1007/s00018-009-0036-0

Estrada, N., Núñez-Vázquez, E. J., Palacios, A., Ascencio, F., Guzmán-Villanueva, L., & Contreras, R. G. (2021). In vitro evaluation of programmed cell death in the immune system of Pacific oyster Crassostrea Gigas by the effect of marine toxins. *Frontiers in Immunology, 12*. https://doi.org/10.3389/fimmu.2021.634497

Federico, S., Pozzetti, L., Papa, A., Carullo, G., Gemma, S., Butini, S., Campiani, G., & Relitti, N. (2020). Modulation of the innate immune response by targeting toll-like receptors: A perspective on their agonists and antagonists. *Journal of Medicinal Chemistry, 63*(22), 13466–13513. https://doi.org/10.1021/acs.jmedchem.0c01049

Fernando, D. D., & Fischer, K. (2020). Proteases and Pseudoproteases in parasitic arthropods of clinical importance. *The FEBS Journal, 287*(19), 4284–4299. https://doi.org/10.1111/febs.15546

García-Carnero, L. C., Martínez-Álvarez, J. A., Salazar-García, L. M., Lozoya-Pérez, N. E., González-Hernández, S. E., & Tamez-Castrellón, A. K. (2020). Recognition of fungal components by the host immune system. *Current Protein and Peptide Science, 21*(3), 245–264. https://doi.org/10.2174/1389203721666191231105546

Ghosh, S. (2018). Environmental pollutants, pathogens and immune system in earthworms. *Environmental Science and Pollution Research, 25*(7), 6196–6208. https://doi.org/10.1007/s11356-017-1167-8

Gómez, X., Sanon, S., Zambrano, K., Asquel, S., Bassantes, M., Morales, J. E., Otáñez, G., et al. (2021). Key points for the development of antioxidant cocktails to prevent cellular stress and damage caused by reactive oxygen species (ROS) during manned space missions. *npj Microgravity, 7*(1), 35. https://doi.org/10.1038/s41526-021-00162-8

Goyal, S., Castrillón-Betancur, J. C., Klaile, E., & Slevogt, H. (2018). The interaction of human pathogenic fungi with C-type lectin receptors. *Frontiers in Immunology, 9.* https://doi.org/10.3389/fimmu.2018.01261

Gundersen-Rindal, D., Dupuy, C., Huguet, E., Jean-Michel, & Drezen. (2013). Parasitoid polydnaviruses: Evolution, pathology and applications: Dedicated to the memory of Nancy E. Beckage. *Biocontrol Science and Technology, 23*(1), 1–61. https://doi.org/10.1080/09583157.2012.731497

Gupta, S., & Yadav, S. (2016). Immuno-defense strategy in earthworms: A review article. *International Journal of Current Microbiology and Applied Sciences, 5*(4), 1022–1035. https://doi.org/10.20546/ijcmas.2016.504.117

Hackett, C. J. (2003). Innate immune activation as a broad-Spectrum biodefense strategy: Prospects and research challenges. *Journal of Allergy and Clinical Immunology, 112*(4), 686–694. https://doi.org/10.1016/S0091-6749(03)02025-6

Hambrook, J. R., & Hanington, P. C. (2021). Immune evasion strategies of Schistosomes. *Frontiers in Immunology, 11.* https://doi.org/10.3389/fimmu.2020.624178

Hassan, S.-u., Donia, A., Sial, U., Zhang, X., & Bokhari, H. (2020). Glycoprotein- and lectin-based approaches for detection of pathogens. *Pathogens, 9*(9), 694. https://doi.org/10.3390/pathogens9090694

Hedstrom, L. (2002). Serine protease mechanism and specificity. *Chemical Reviews, 102*(12), 4501–4524. https://doi.org/10.1021/cr000033x

Holman, J. D., Burnett, K. G., & Burnett, L. E. (2004). Effects of hypercapnic hypoxia on the clearance of vibrio Campbellii in the Atlantic Blue Crab, Callinectes Sapidus Rathbun. *The Biological Bulletin, 206*(3), 188–196. https://doi.org/10.2307/1543642

Holmblad, T., & Söderhäll, K. (1999). Cell adhesion molecules and antioxidative enzymes in a crustacean, possible role in immunity. *Aquaculture, 172*(1), 111–123. https://doi.org/10.1016/S0044-8486(98)00446-3

Islam, M. N., Rauf, A., Fahad, F. I., Emran, T. B., Mitra, S., Olatunde, A., Shariati, M. A., et al. (2022). Superoxide dismutase: An updated review on its health benefits and industrial applications. *Critical Reviews in Food Science and Nutrition, 62*(26), 7282–7300. https://doi.org/10.1080/10408398.2021.1913400

Johnson, V. J. M., Thiraviam, A. G. P., Chellathangam, A., Ramaswamy, V. D., & Rajam, B. R. M. (2022). Lectins in penaeid shrimps: purification, characterization, and biological significance. In P. Elumalai, B. Vaseeharan, & S. Lakshmi (Eds.), *Aquatic lectins: Immune defense, biological recognition and molecular advancements* (pp. 125–167). Springer Nature. https://doi.org/10.1007/978-981-19-0432-5_7

Kawabata, S., Koshiba, T., & Shibata, T. (2009). The lipopolysaccharide-activated innate immune response network of the horseshoe crab. *Invertebrate Survival Journal, 6*(1), 59–77.

Köhlerová, P., Beschin, A., Šilerová, M., De Baetselier, P., & Bilej, M. (2004). Effect of experimental microbial challenge on the expression of defense molecules in *Eisenia Foetida* earthworm. *Developmental & Comparative Immunology, 28*(7), 701–711. https://doi.org/10.1016/j.dci.2004.01.001

Kuo, C.-J., Hansen, M., Emily, & Troemel. (2018). Autophagy and innate immunity: Insights from invertebrate model organisms. *Autophagy, 14*(2), 233–242. https://doi.org/10.1080/15548627.2017.1389824

Lai, Y., & Gallo, R. L. (2009). AMPed up immunity: How antimicrobial peptides have multiple roles in immune defense. *Trends in Immunology, 30*(3), 131–141. https://doi.org/10.1016/j.it.2008.12.003

Leusmann, S., Ménová, P., Shanin, E., Titz, A., & Rademacher, C. (2023). Glycomimetics for the inhibition and modulation of lectins. *Chemical Society Reviews, 52*(11), 3663–3740. https://doi.org/10.1039/D2CS00954D

Li, D., & Wu, M. (2021). Pattern recognition receptors in health and diseases. *Signal Transduction and Targeted Therapy, 6*(1), 1–24. https://doi.org/10.1038/s41392-021-00687-0

Li, H., Zhang, H., Jiang, S., Wang, W., Xin, L., Wang, H., Wang, L., & Song, L. (2015). A single-CRD C-type lectin from oyster *Crassostrea Gigas* mediates immune recognition and pathogen elimination with a potential role in the activation of complement system. *Fish & Shellfish Immunology, 44*(2), 566–575. https://doi.org/10.1016/j.fsi.2015.03.011

Li, Y., Song, X., Wang, W., Wang, L., Yi, Q., Jiang, S., Jia, Z., et al. (2017). The hematopoiesis in gill and its role in the immune response of Pacific oyster *Crassostrea Gigas* against secondary challenge with *Vibrio Splendidus. Developmental & Comparative Immunology, 71*, 59–69. https://doi.org/10.1016/j.dci.2017.01.024

Liegeois, S., & Ferrandon, D. (2022). Sensing microbial infections in the drosophila melanogaster genetic model organism. *Immunogenetics, 74*(1), 35–62. https://doi.org/10.1007/s00251-021-01239-0

Lin, Y.-C., Vaseeharan, B., & Chen, J.-C. (2008). Identification and phylogenetic analysis on lipopolysaccharide and β-1,3-Glucan binding protein (LGBP) of Kuruma shrimp *Marsupenaeus Japonicus. Developmental & Comparative Immunology, 32*(11), 1260–1269. https://doi.org/10.1016/j.dci.2008.05.003

Liu, D. W., Chen, Z. W., & Xu, H. Z. (2008). Effects of leucine-enkephalin on catalase activity and hydrogen peroxide levels in the haemolymph of the Pacific Oyster (Crassostrea gigas). *Molecules, 13*(4), 864–870.

Liu, C., Zhang, T., Wang, L., Wang, M., Wang, W., Jia, Z., Jiang, S., & Song, L. (2016). The modulation of extracellular superoxide dismutase in the specifically enhanced cellular immune response against secondary challenge of *vibrio Splendidus* in Pacific oyster (*Crassostrea Gigas*). *Developmental & Comparative Immunology, 63*, 163–170. https://doi.org/10.1016/j.dci.2016.06.002

Luo, Y., & Song, Y. (2021). Mechanism of antimicrobial peptides: Antimicrobial, anti-inflammatory and antibiofilm activities. *International Journal of Molecular Sciences, 22*(21), 11401. https://doi.org/10.3390/ijms222111401

Mahla, R. S., Madhava Reddy, C., Prasad, D., & Kumar, H. (2013). Sweeten PAMPs: Role of sugar complexed PAMPs in innate immunity and vaccine biology. *Frontiers in Immunology, 4*. https://doi.org/10.3389/fimmu.2013.00248

Mao, F., Lin, Y., He, Z., Li, J., Xiang, Z., Zhang, Y., & Ziniu, Y. (2018). Dual roles of cystatin a in the immune defense of the Pacific oyster, *Crassostrea Gigas. Fish & Shellfish Immunology, 75*, 190–197. https://doi.org/10.1016/j.fsi.2018.01.041

Mayer, S., Raulf, M.-K., & Lepenies, B. (2017). C-type lectins: Their network and roles in pathogen recognition and immunity. *Histochemistry and Cell Biology, 147*(2), 223–237. https://doi.org/10.1007/s00418-016-1523-7

Meister, M., Hetru, C., & Hoffmann, J. A. (2000). The antimicrobial host defense of drosophila. In L. Du Pasquier & G. W. Litman (Eds.), *Origin and evolution of the vertebrate immune system* (pp. 17–36). Springer. https://doi.org/10.1007/978-3-642-59674-2_2

Mukwevho, E., Ferreira, Z., & Ayeleso, A. (2014). Potential role of sulfur-containing antioxidant systems in highly oxidative environments. *Molecules, 19*(12), 19376–19389. https://doi.org/10.3390/molecules191219376

Narayanan, K. B. (2025). Enzyme-based anti-inflammatory therapeutics for inflammatory diseases. *Pharmaceutics, 17*(5), 606. https://doi.org/10.3390/pharmaceutics17050606

Netto, L. E. S., & Antunes, F. (2016). The roles of peroxiredoxin and thioredoxin in hydrogen peroxide sensing and in signal transduction. *Molecules and Cells, 39*(1), 65–71. https://doi.org/10.14348/molcells.2016.2349

Oakley, C. A., Durand, E., Wilkinson, S. P., Peng, L., Weis, V. M., Grossman, A. R., & Davy, S. K. (2017). Thermal shock induces host proteostasis disruption and endoplasmic reticulum stress in the model symbiotic cnidarian Aiptasia. *Journal of Proteome Research, 16*(6), 2121–2134.

Otto, H.-H., & Schirmeister, T. (1997). Cysteine proteases and their inhibitors. *Chemical Reviews, 97*(1), 133–172. https://doi.org/10.1021/cr950025u

Oyinloye, B. E., Adenowo, A. F., & Kappo, A. P. (2015). Reactive oxygen species, apoptosis, antimicrobial peptides and human inflammatory diseases. *Pharmaceuticals, 8*(2), 151–175. https://doi.org/10.3390/ph8020151

Patrnogic, J. (2014). *'Serine proteases and serine protease homologs: genetic analysis of their involvement in immune response activation in Drosophila'. phdthesis*. Université de Strasbourg. https://theses.hal.science/tel-01595747 (June 18, 2025)

Pirtskhalava, M., Vishnepolsky, B., Grigolava, M., & Managadze, G. (2021). Physicochemical features and peculiarities of interaction of AMP with the membrane. *Pharmaceuticals, 14*(5), 471. https://doi.org/10.3390/ph14050471

Potęga, A. (2022). Glutathione-mediated conjugation of anticancer drugs: An overview of reaction mechanisms and biological significance for drug detoxification and bioactivation. *Molecules, 27*(16), 5252. https://doi.org/10.3390/molecules27165252

Qiao, Z., Li, P., Tan, J., Peng, C., Zhang, F., Zhang, W., & Jiang, X. (2022). Oxidative stress and detoxification mechanisms of earthworms (*Eisenia Fetida*) after exposure to flupyradifurone in a soil-earthworm system. *Journal of Environmental Management, 322*, 115989. https://doi.org/10.1016/j.jenvman.2022.115989

Ragland, S. A., & Criss, A. K. (2017). From bacterial killing to immune modulation: Recent insights into the functions of lysozyme. *PLoS Pathogens, 13*(9), e1006512. https://doi.org/10.1371/journal.ppat.1006512

Rahnamaeian, M. (2011). Antimicrobial peptides: Modes of mechanism, modulation of defense responses. *Plant Signaling & Behavior, 6*(9), 1325–1332. https://doi.org/10.4161/psb.6.9.16319

Ramesh, S., Govender, T., Kruger, H. G., de la Torre, B. G., & Albericio, F. (2016). Short antimicrobial peptides (SAMPs) as a class of extraordinary promising therapeutic agents. *Journal of Peptide Science, 22*(7), 438–451. https://doi.org/10.1002/psc.2894

Reichhart, J. M., Gubb, D., & Leclerc, V. (2011). The *drosophila* serpins. In J. C. Whisstock & P. I. Bird (Eds.), *Methods in enzymology, biology of serpins* (pp. 205–225). Academic Press. https://doi.org/10.1016/B978-0-12-386471-0.00011-0

Ren, L., Liao, M., Ruixue, H., Feng, G., Chen, N., & Zemao, G. (2024). *Pc*LGBP, a pattern recognition receptor from *Procambarus Clarkii* confers broad-spectrum disease resistance by activating ProPO system and AMPs related pathways. *Aquaculture, 585*, 740673. https://doi.org/10.1016/j.aquaculture.2024.740673

Rosilan, N. F., Waiho, K., Fazhan, H., Sung, Y. Y., Nor, S. A. M., Muhammad, N. A. N., Mohamed-Hussein, Z.-A., & Afiqah-Aleng, N. (2023). Protein-protein interaction network analysis on the Whiteleg shrimp *Penaeus Vannamei* and *Vibrio Parahaemolyticus* host-pathogen relationship reveals possible proteins and pathways involved during infection. *Aquaculture Reports, 30*, 101583. https://doi.org/10.1016/j.aqrep.2023.101583

Roy, S., Bossier, P., Norouzitallab, P., & Vanrompay, D. (2020). Trained immunity and perspectives for shrimp aquaculture. *Reviews in Aquaculture, 12*(4), 2351–2370. https://doi.org/10.1111/raq.12438

Royet, J. (2004). Drosophila melanogaster innate immunity: An emerging role for peptidoglycan recognition proteins in bacteria detection. *Cellular and Molecular Life Sciences CMLS, 61*(5), 537–546. https://doi.org/10.1007/s00018-003-3243-0

Sacchi, S., Malagoli, D., & Franchi, N. (2024). The invertebrate immunocyte: A complex and versatile model for immunological, developmental, and environmental research. *Cells, 13*(24), 2106. https://doi.org/10.3390/cells13242106

Saeed, K., Riaz, S., Adil, A., Nawaz, I., Kamran-U, S., Naqvi, H., Baig, A., Ali, M., et al. (2023). Characterization of alkaline metalloprotease isolated from halophilic bacterium *Bacillus cereus* and its applications in various industrial processes. *Anais da Academia Brasileira de Ciências, 95*, e20230014. https://doi.org/10.1590/0001-3765202320230014

Schmitt, P., Wilmes, M., Pugnière, M., Aumelas, A., Bachère, E., Sahl, H.-G., Schneider, T., & Destoumieux-Garzón, D. (2010). Insight into invertebrate Defensin mechanism of action:

Oyster defensins inhibit peptidoglycan biosynthesis by binding to lipid II*. *Journal of Biological Chemistry, 285*(38), 29208–29216. https://doi.org/10.1074/jbc.M110.143388

Sevilla, F., Camejo, D., Ortiz-Espín, A., Calderón, A., Lázaro, J. J., & Jiménez, A. (2015). The thioredoxin/peroxiredoxin/sulfiredoxin system: Current overview on its redox function in plants and regulation by reactive oxygen and nitrogen species. *Journal of Experimental Botany, 66*(10), 2945–2955. https://doi.org/10.1093/jxb/erv146

Seyfi, R., Kahaki, F. A., Ebrahimi, T., Montazersaheb, S., Eyvazi, S., Babaeipour, V., & Tarhriz, V. (2020). Antimicrobial peptides (AMPs): Roles, functions and mechanism of action. *International Journal of Peptide Research and Therapeutics, 26*(3), 1451–1463. https://doi.org/10.1007/s10989-019-09946-9

Shaka, M., Arias-Rojas, A., Hrdina, A., Frahm, D., & Iatsenko, I. (2022). Lipopolysaccharide-mediated resistance to host antimicrobial peptides and hemocyte-derived reactive-oxygen species are the major Providencia Alcalifaciens virulence factors in drosophila melanogaster. *PLoS Pathogens, 18*(9), e1010825. https://doi.org/10.1371/journal.ppat.1010825

Shankar, R., Upadhyay, P. K., & Kumar, M. (2021). Protease enzymes: Highlights on potential of proteases as therapeutics agents. *International Journal of Peptide Research and Therapeutics, 27*(2), 1281–1296. https://doi.org/10.1007/s10989-021-10167-2

Shekhova, E. (2020). Mitochondrial reactive oxygen species as major effectors of antimicrobial immunity. *PLoS Pathogens, 16*(5), e1008470. https://doi.org/10.1371/journal.ppat.1008470

Singh, R. P., & Bhardwaj, A. (2023). β-Glucans: A potential source for maintaining gut microbiota and the immune system. *Frontiers in Nutrition, 10*. https://doi.org/10.3389/fnut.2023.1143682

Singh, S., Singh, A., Kumar, S., Mittal, P., & Singh, I. K. (2020). Protease inhibitors: Recent advancement in its usage as a potential biocontrol agent for insect pest management. *Insect Science, 27*(2), 186–201. https://doi.org/10.1111/1744-7917.12641

Škanta, F. (2016). *Recognition of microbial patterns in earthworms.* https://dspace.cuni.cz/handle/20.500.11956/82419 (June 18, 2025).

Söderhäll, K. (2011). *Invertebrate immunity.* Springer Science & Business Media.

Söderhäll, K., & Cerenius, L. (1998). Role of the prophenoloxidase-activating system in invertebrate immunity. *Current Opinion in Immunology, 10*(1), 23–28. https://doi.org/10.1016/S0952-7915(98)80026-5

Stein, E. A., Younai, S., & Cooper, E. L. (1986). Bacterial agglutinins of the earthworm, *Lumbricus Terrestris. Comparative Biochemistry and Physiology Part B: Comparative Biochemistry, 84*(3), 409–415. https://doi.org/10.1016/0305-0491(86)90099-4

Steiner, H. (2004). Peptidoglycan recognition proteins: On and off switches for innate immunity. *Immunological Reviews, 198*(1), 83–96. https://doi.org/10.1111/j.0105-2896.2004.0120.x

Sukhithasri, V., Nisha, N., Lalitha Biswas, V., Kumar, A., & Biswas, R. (2013). Innate immune recognition of microbial cell wall components and microbial strategies to evade such recognitions. *Microbiological Research, 168*(7), 396–406. https://doi.org/10.1016/j.micres.2013.02.005

Sun, L., Jia, M., Zhu, K., Hao, Z., Shen, J., & Wang, S. (2024). The efficacy of cecropin against multidrug-resistant bacteria is linked to the destabilization of outer membrane structure LPS of gram-negative bacteria. *Probiotics and Antimicrobial Proteins.* https://doi.org/10.1007/s12602-024-10424-y

Tavassolifar, M. J., Vodjgani, M., Salehi, Z., & Izad, M. (2020). The influence of reactive oxygen species in the immune system and pathogenesis of multiple sclerosis. *Autoimmune Diseases, 2020*(1), 5793817. https://doi.org/10.1155/2020/5793817

Telfer, W.H, & Kunkel, J. G. (n.d.). *The function and evolution of insect storage hexamers*

Terwilliger, N. B. (2015). Oxygen transport proteins in Crustacea: Hemocyanin and hemoglobin. *Physiology, 4*, 359–390.

Udompetcharaporn, A., Junkunlo, K., Senapin, S., Roytrakul, S., Flegel, T. W., & Sritunyalucksana, K. (2014). Identification and characterization of a QM protein as a possible peptidoglycan recognition protein (PGRP) from the Giant Tiger shrimp *Penaeus Monodon. Developmental & Comparative Immunology, 46*(2), 146–154. https://doi.org/10.1016/j.dci.2014.04.003

Vasta, G. R., & Wang, J.-X. (2020). Galectin-mediated immune recognition: Opsonic roles with contrasting outcomes in selected shrimp and bivalve mollusk species. *Developmental & Comparative Immunology, 110*, 103721. https://doi.org/10.1016/j.dci.2020.103721

Veillard, F., Troxler, L., & Reichhart, J.-M. (2016). *Drosophila melanogaster* clip-domain serine proteases: Structure, function and regulation. *Biochimie, 122*, 255–269. https://doi.org/10.1016/j.biochi.2015.10.007

Verma, S., Dixit, R., & Pandey, K. C. (2016). Cysteine proteases: Modes of activation and future prospects as pharmacological targets. *Frontiers in Pharmacology, 7*. https://doi.org/10.3389/fphar.2016.00107

Wang, X.-W., Ji-Dong, X., Zhao, X.-F., Vasta, G. R., & Wang, J.-X. (2014). A shrimp C-type lectin inhibits proliferation of the hemolymph microbiota by maintaining the expression of antimicrobial peptides. *Journal of Biological Chemistry, 289*(17), 11779–11790.

Wei, L., Yang, Y., Zhou, Y., Li, M., Yang, H., Lixian, M., Qian, Q., Jing, W., & Wei, X. (2018). Anti-inflammatory activities of Aedes Aegypti Cecropins and their protection against murine endotoxin shock. *Parasites & Vectors, 11*(1), 470. https://doi.org/10.1186/s13071-018-3000-8

Wojda, I., Cytryńska, M., Zdybicka-Barabas, A., & Kordaczuk, J. (2020). Insect defense proteins and peptides. In U. Hoeger & J. Robin Harris (Eds.), *Vertebrate and invertebrate respiratory proteins, lipoproteins and other body fluid proteins* (pp. 81–121). Springer International Publishing. https://doi.org/10.1007/978-3-030-41769-7_4

Wongpanya, R., Sengprasert, P., Amparyup, P., & Tassanakajon, A. (2017). A novel C-type lectin in the black Tiger shrimp *Penaeus Monodon* functions as a pattern recognition receptor by binding and causing bacterial agglutination. *Fish & Shellfish Immunology, 60*, 103–113. https://doi.org/10.1016/j.fsi.2016.11.042

Xiang, Z., Fufa, Q., Wang, F., Li, J., Zhang, Y., & Ziniu, Y. (2014). Characteristic and functional analysis of a Ficolin-like protein from the oyster *Crassostrea Hongkongensis*. *Fish & Shellfish Immunology, 40*(2), 514–523. https://doi.org/10.1016/j.fsi.2014.08.006

Xue, Q. (2019). Pathogen proteases and host protease inhibitors in molluscan infectious diseases. *Journal of Invertebrate Pathology, 166*, 107214. https://doi.org/10.1016/j.jip.2019.107214

Yang, N., Matthew, M. A., & Yao, C. (2023). Roles of cysteine proteases in biology and pathogenesis of parasites. *Microorganisms, 11*(6), 1397. https://doi.org/10.3390/microorganisms11061397

Zannella, C., Mosca, F., Mariani, F., Franci, G., Folliero, V., Galdiero, M., Tiscar, P. G., & Galdiero, M. (2017). Microbial diseases of bivalve mollusks: Infections, immunology and antimicrobial defense. *Marine Drugs, 15*(6), 182. https://doi.org/10.3390/md15060182

Zdybicka-Barabas, A., Stączek, S., Kunat-Budzyńska, M., & Cytryńska, M. (2025). Innate immunity in insects: The lights and shadows of phenoloxidase system activation. *International Journal of Molecular Sciences, 26*(3), 1320. https://doi.org/10.3390/ijms26031320

Zhao, H. W., Zhou, D., & Haddad, G. G. (2011). Antimicrobial peptides increase tolerance to oxidant stress in drosophila melanogaster*. *Journal of Biological Chemistry, 286*(8), 6211–6218. https://doi.org/10.1074/jbc.M110.181206

Zhao, L., Niu, J., Feng, D., Wang, X., & Zhang, R. (2023). Immune functions of pattern recognition receptors in Lepidoptera. *Frontiers in Immunology, 14*. https://doi.org/10.3389/fimmu.2023.1203061

Zheng, X., Duan, Y., Dong, H., & Zhang, J. (2017). Effects of dietary lactobacillus Plantarum in different treatments on growth performance and immune gene expression of white shrimp Litopenaeus Vannamei under normal condition and stress of acute low salinity. *Fish & Shellfish Immunology, 62*, 195–201.

Zhou, L., Meng, G., Zhu, L., Ma, L., & Chen, K. (2024). Insect antimicrobial peptides as guardians of immunity and beyond: A review. *International Journal of Molecular Sciences, 25*(7), 3835. https://doi.org/10.3390/ijms25073835

Zhu, Z., Deng, X., Xie, W., Li, H., Li, Y., & Deng, Z. (2024). Pharmacological effects of bioactive agents in earthworm extract: A comprehensive review. *Animal Models and Experimental Medicine, 7*(5), 653–672. https://doi.org/10.1002/ame2.12465

Chapter 4
Advanced Molecular Biology Techniques in Innate Immunity

4.1 Invertebrate Immune Genes Regulatory Mechanisms Through the Operon Concept

While lacking adaptive immunity, invertebrates maintain a very efficient innate immune system to combat an array of pathogens. Detecting and eliminating harmful microorganisms is accomplished by various proteins and signaling pathways that mediate this innate immunity. The operon concept is one of the major regulatory mechanisms that govern these immune responses in invertebrates. The first described operon model consists of the regulation of a cluster of genes whose expression is co-regulated and co-expressed under the control of a single promoter and regulatory elements of prokaryotic organisms. All of these regulatory systems are found in invertebrate immunity in the context of expression of immune-related genes, specifically antimicrobial peptides (AMPs), pattern recognition receptors (PRRs) and the set of other immune effector molecules. The operon model is critical to quickly and efficiently activate immune responses to a pathogen detection. The model puts together the expression of multiple immune genes, which enables invertebrates to rapidly build a broad immune defense. Here, immune genes are considered as an example within the concept of operon to review how operon influences immune gene regulation in invertebrates using arthropods, mollusks, annelids, and other phyla in particular (Table 4.1).

4.2 Operon Concept in Invertebrate Immunity

In invertebrates, regulation of multiple immune-related gene by a single regulatory element is the operon concept. Upon pathogen detection the regulatory proteins bind to the operon's promoter region initiating the transcription and concurrent

© The Author(s), under exclusive license to Springer Nature Singapore Pte Ltd. 2025
S. Jeyachandran, B. A. Venmathi Maran, *Invertebrate Immunology*,
https://doi.org/10.1007/978-981-95-1549-3_4

Table 4.1 Examples of immune gene operons in invertebrates

Organism	Operon components	Immune function	Type of regulation	Reference
D. melanogaster	Drosomycin, Cecropin, Attacin	AMP production	Toll & Imd pathway	Lemaitre et al. (1996)
C. gigas	Lectin, Lysozyme, Defensin	Bacterial resistance	PRR activation	Bachère et al. (2004)
P. monodon	ProPO, Penaeidins, Lysozyme	Viral & bacterial immunity	Pathogen-induced	Zhang et al. (2013)
L. terrestris	ProPO genes	Melanization, Encapsulation	Coordinated regulation	Söderhäll and Cerenius (1998)
S. mansoni	Surface antigens, suppressors	Immune evasion	Host-pathogen interface	Cioli et al. (1995)
C. vicina	Melanization operon	Encapsulation	Co-expression	Cerenius et al. (2004)
H. verbana	Hemocyte lectin complex	Bacterial defense	Localized activation	N/A
B. glabrata	FREP clusters	Parasite recognition	Immune memory-like	Zhang et al. (2020)
M. galloprovincialis	Mytilin & defensin-like genes	Broad-spectrum AMP	Operonic induction	N/A
C. elegans	FIPR genes	Immune signaling	PRR-activated	Schulenburg et al. (2007)

expression of several immune genes. By co-regulating this immune system, it is ensured that the immune system responds well to infections. In these operons, genes encoding pattern recognition, signaling, and antimicrobial peptides are often coexpressed among these operons for quick recognition of pathogens and triggering a defense response. Regulation of immune genes via operon is very well conserved across divergent invertebrates. In general, such transcription factors are components of well-known immune signal transduction pathways (e.g., Toll and Imd pathways), which, when activated by infection, cause antimicrobial peptide (AMP) expression. Rapid and coordinated immune response is achieved using operon concept, which means the expression of these AMPs and other immune genes can quickly be expressed and expressed simultaneously. In *Drosophila melanogaster*, one of the best studied examples for the regulation of immune genes through operons, AMPs are expressed during infection. *Drosophila* immune system mounts expression of several AMPs such as drosomycins, cecropins, and attacins, that are activated through activation of the Toll and Imd signaling pathways. AMP genes are often found clustered together in operons such that they are expressed simultaneously when the immune response is turned on and results in the transcription of the AMP genes. When Drosophila is infected with bacteria or fungi the Toll receptor is activated and that activates the NF-κB family of transcription factors, Dorsal and Relish for example. These transcription factors bind to the promoter regions of immune operons and cause the coordinated expression of multiple AMPs. The operon nature of these immune genes means that they are expressed together to provide a broader spectrum defense against invading pathogen (Lemaitre et al., 1996).

Operons regulate immune genes in the *Crassostrea gigas* through response to bacterial infections, especially that of *Vibrio* species. Bacteria are detected by PRRs including lectins performing pattern recognition based on pathogen associated molecular patterns (PAMPs) on bacterial surface. When recognized by these receptors, the latter will activate immune signaling pathways that will cause the expression of antimicrobial peptides (AMPs), such as defensins or lysozymes. AMP genes are often organized in operons so that they can be expressed coordinately. Activation of immune operons in hemocytes and other immune tissues of *C. gigas* leads to the production of multiple AMPs simultaneously in order to promote a resistance to the infection. This is very important for the defense of the oyster against pathogens in a constantly wet environment, where the oyster is constantly faced with attacking microbes (Bachère et al., 2004). WSSV elicits the immune response in shrimp (*Penaeus monodon*) through the induction of immune-related genes expression such as antimicrobial peptides (AMPs), prophenoloxidase (proPO) and other immune proteins. These immune genes are often a part of operons and they therefore ensure expression of these genes together, once infected. Pathogen recognition and encapsulation rely on the prophenoloxidase system. Upon recognition of viral pathogens by pattern recognition receptors, shrimp activate immune operons by triggering immune signaling pathways. The prompt and effective activation of the immune response, which saves shrimp from viral infections, is ensured by the fact that the expression of proPO and other immune genes takes place within operons (Zhang et al., 2013).

The immune response in such earthworms as *Lumbricus terrestris* is regulated by operons controlling the expression of the genes recognizing pathogen, enclosing pathogens in capsules or generating antimicrobial defense. Operons, which contain the genes of the prophenoloxidase (proPO) system, are a major component of earthworm immunity (Fig. 4.1). Bacteria infected into *L. terrestis* trigger the proPO pathway, which in turn activates a protein, which results in the formation of melanin encapsulating and thus killing the pathogens. Operons that regulate proPO expression are required to activate multiple proPO components of an immune response together for a coordinated and efficient response to microbe invasion (Söderhäll & Cerenius, 1998); in parasitic flatworms such as *Schistosoma mansoni* operons controlling immune genes are involved in immune evasion and modulation. Finally, these operons regulate the expression of surface proteins that aid the parasite to evade the host immune system. Immune evasion is critical for survival of the parasite Schistosoma, and operon regulation of surface proteins is stringently regulated in this parasite. Schistosoma mansoni utilizes their surface proteins to interact with host immune receptors, blocking activation of immune pathways and thus ensuring survival of the parasite inside its host. Operons that control the expression of these proteins relate to immune modulation and host–parasite interactions are regulated. Similarly, in the blowfly *Calliphora vicina*, coordinate production of prophenoloxidase (proPO) and other proteins in the melanization pathway is involved in the

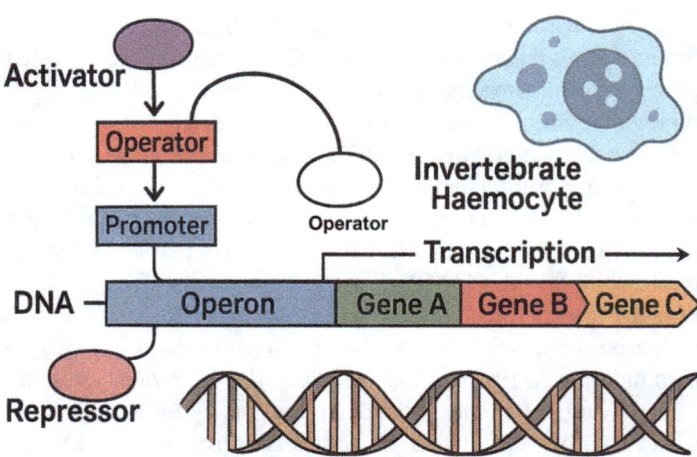

Fig. 4.1 Illustrates genetic organization and transcriptional regulation of immune-related operons in hemocytes. Shows control elements and gene clusters responsible for coordinated immune responses

functioning of immune operon. Because the melanization pathway results in the production of melanin surrounding and inactivating the invading pathogen, this pathway is essential in pathogen neutralization. Through this operon-based regulation the melanization response is efficiently activated and can be considered as very effective defense against pathogens.

4.3 Significance of Operon Regulation in Invertebrate Immunity

It is important for the concept of an operon in order to coordinate and efficiently regulate the immune responses in invertebrate immunity. They group together the related immune genes into an operon to be expressed together, because many immune genes serve together to fight pathogens, so should be expressed together. The co-regulation of this immune system production guarantees that the immune system is ready and can rapidly respond to the interrogation of an infection and produce from one of a very wide range of immune molecules required for such rapid action including the presence of antimicrobial peptides, cytokines or enzymes that contribute to the disruption of capsule or melanization of the pathogen. This also provides an efficient mechanism of fine tuning of immune responses. Upon pathogen detection, the regulatory proteins bind to the immune operon promoter regions and thereby up-regulate multiple genes for immune defense. Regulation is coordinated so that pathogens can be recognized and eliminated at multiple levels from recognition to elimination, and this in turn makes the operon concept a very powerful framework for understanding regulation of invertebrate immune genes. Through regulating the expression of multiple immune genes simultaneously, operons facilitate invertebrates either to mount synchronized and rapid or efficient immune responses coinciding with the threats of pathogens. Operon concept is used to define the immune gene expression in different invertebrate species, for example *D. melanogaster, C. gigas, P. monodon,* and *L. terrestris* as shown from the examples in this review. The co-regulation is essential for immune homeostasis and for an immune response against infections. This model of operon sheds light onto molecular mechanisms at the bottom of the innate immunity in invertebrates, and opens new directions of research in the field of the immune gene regulation and in therapeutic application with respect to the health of invertebrates.

4.4 Gene Expression Analysis Using qPCR in Invertebrates: Methodology, Applications, and Examples

The study of the regulatory control of genes mediating all these biological processes such as immune responses, development, and responses to environmental adaptation, is powerfully facilitated by gene expression analysis. One of the most used technique of gene expression analysis is quantitative Polymerase Chain Reaction (qPCR) or real time PCR. In the use of qPCR for measuring the expression of immune-related genes, as has been in practice for invertebrate immunology, the quantification of precise levels of gene expression of a particular gene in specific tissues or cells has allowed for inferring effects of regulation under such and such conditions. This method allows researchers to quantify the relative expression levels of genes involved in the innate immune response in pattern recognition receptors (PRRs) of the innate immune response, antimicrobial peptides (AMPs) and signaling proteins. The methodology of qPCR is reviewed and its application in invertebrate immunity with several examples from the different phyla of the invertebrates to show its use to investigate the invertebrate immune response. Using of qPCR in *D. melanogaster* to measure the expression of the immune-correlated genes, including the genes at Toll and IMD pathway is very wide. When *Pseudomonas aeruginosa* gets activated, it activates these pathways. Expression of qPCR is upregulated under infection of drosomycin (defensin, an antimicrobial peptide in drosomycin) (Kuraishi et al., 2011). In addition, qPCR is used to investigate the activation of immune signaling pathways activated on bacterial infection due to common occurrence of bacterial infection of larvae during embryogenesis by studying genes encoding pattern recognition receptors (PRRs) such as Toll receptors and NF-κB transcription factor Relish. The immune response of AMP genes is characterized with significant upregulations in the immune response of infected Drosophila larvae and qPCR assay can be used to determine its characterization. qPCR has been used in this shellfish for the study of the immune response to bacterial infection in the *C. gigas,* and the most common pathogens involved are *Vibrio* spp. It has quantitated expression of defensin and lysozyme coding genes (antimicrobial peptides) and expression of hemocyte lytic immune proteins. In turn, the pathogen exposure up regulated these immune genes as a rapid immune response (Bachère et al., 2004). Defensin RNAs were quantified by qPCR after exposure of oysters to oysters to Vibrio species (Table 4.2).

The study of the immune response to the shrimp like *Penaeus monodon* of *Vibrio harveyi* bacterial pathogen by qPCR. This included measurement of immune gene expression (penaeidins, prophenoloxidase (proPO) and lysozymes) that have shown shrimp upregulates its immune protein expression in response to bacterial infection, and quantification of changes in gene expression using qPCR, which allowed us to learn about molecular immune defense mechanisms in crustaceans (Zhang et al., 2013). qPCR was used to measure the levels of penaeidins and prophenoloxidase in infected shrimp in order to understand immune activation in response to *Vibrio* infection. Immune genes including immunity to bacterial pathogens have been assayed in earthworms by qPCR, such as *L. terrestris*. With qPCR they used to compare the expression of immune genes in earthworms exposed to pathogens and to explore how earthworms react to pathogens by quantifying the expression of genes encoding antimicrobial peptides and other enzymes such as lysozymes and proteases that destroy pathogens as these genes are directly involved in clearing pathogens, and Söderhäll and Cerenius (1998) find that some immune genes are significantly up regulated following bacterial infection. qPCR was applied to the study of host immune response to the parasitic worm *Schistosoma mansoni* and the results were also confirmed using qPCR, and *L. terrestris* also mounted a strong immune response to both compounds. Immune response to *S. mansoni* infection is due to both host and parasite's immune pathway activation. Researchers have used

Table 4.2 Invertebrate immune genes analyzed by qPCR

Species	Target gene	Immune role	Infection model	Reference
D. melanogaster	Drosomycin	Antifungal AMP	*P. aeruginosa*	Kuraishi et al. (2011)
C. gigas	Lysozyme	Bacterial lysis	*Vibrio* spp.	Bachère et al. (2004)
P. monodon	ProPO	Melanization	*V. harveyi*	Zhang et al. (2013)
L. terrestris	Lysozyme, Proteases	Pathogen digestion	Mixed bacterial	Söderhäll and Cerenius (1998)
S. mansoni	Cytokines	Host modulation	Host–parasite model	
B. mori	Cecropin B	Viral defense	BmNPV infection	
H. verbana	Lectins	Hemocyte activation	Bacterial challenge	
C. elegans	SOD-3	ROS defense	Fungal pathogens	
A. millepora	HSP70	Thermal response	Heat stress	
M. sexta	PPO	PO activation	Microbial injection	

qPCR to measure the expression of immune genes in the host, such as cytokines, chemokines, and AMPs, to understand how the host's immune system responds to the parasite. qPCR analysis also helps assess the parasite's ability to modulate the immune response to evade host defense mechanisms. qPCR was used to measure the expression of cytokine genes in the host *during S. mansoni* infection, revealing significant changes in immune gene expression in response to parasitic challenge. One advantage that qPCR has over classical used techniques like Northern blotting or enzyme-linked immunosorbent assays (ELISA) for this kind of research on immune response are: qPCR is possible to quantify gene expression with high sensitivity and specificity and ability to detect low abundance transcript. Conventional PCR lacks the capability of returning quantitative values of genes expression whereas qPCR can allow accurate and quantitative measurement of variations in immune gene expression over time or in relation to any condition. qPCR is applied to a variety of invertebrate species, such marine mollusks (*C. gigas*), (*D. melanogaster*) arthropods, and as such reveals conserved and divergent aspects of the immune response. Being relatively fast compared to RNA-Seq, needing little resources and is excellent for high throughput studies of immune gene expression, qPCR is an indispensable tool in studying innate immunity in invertebrates by the means of gene expression analyses where it allows for accurate quantification of gene expression on the level of transcript and is used by researchers to investigate the molecular mechanisms of pathogen detection, immune tagging, and immune signaling. These mentioned examples in the review illustrate the amount of major qPCR applications in different invertebrate species (*D. melanogaster, C. gigas, P. monodon,* and *L. terrestris*). The results of these studies can be used to help understand the functions and their pathways in the immune genes in protection against bacterial, fungal, and parasitic infections. While qPCR will remain a key method by which our knowledge of invertebrate innate immunity and its control is increased, continued development of the technique will ensure that this is not an outdated technique.

4.5 Invertebrate Immune Pathways Regulatory Mechanisms Through dsRNA, RNAi, and miRNA: Insights Across Phyla

Invertebrates have highly effective innate immune systems, which involve the usage of a variety of molecular and cellular strategies for rapid, nonspecific, and effective defense against pathogens (Table 4.3). In invertebrates, RNA biology is key in the regulation of immune responses; for example, different technologies including use of double stranded RNA (dsRNA), RNA interference (RNAi), and microRNAs (miRNAs) have been developed to understand and modulate immune pathways. These have become very useful to identify and manipulate immune-related genes and define their function in pathogen recognition, immune activation, and an

Table 4.3 Application of RNAi, dsRNA, and miRNA in immune regulation

Tool	Organism	Target gene	Outcome	Reference
RNAi	D. melanogaster	Toll, Imd	AMP reduction	Lemaitre et al. (1996)
RNAi	C. gigas	Lectins	Infection susceptibility ↑	Bachère et al. (2004)
dsRNA	P. monodon	Penaeidins	WSSV susceptibility ↑	Zhang et al. (2013)
dsRNA	D. melanogaster	Viral gene mimic	Vago expression ↑	
miRNA	D. melanogaster	miR-14	AMP overexpression	
miRNA	C. gigas	miR-122	PRR suppression ↓	
RNAi	L. terrestris	PO genes	Reduced encapsulation	
dsRNA	H. verbana	PRRs	Inhibition of hemocyte activity	
miRNA	B. mori	miR-263	Immune modulation	
RNAi	C. elegans	Pathway genes	Reduced stress proteins	

immune response. Here I review how dsRNA, RNAi and miRNA participate in regulating immune pathways in invertebrates by examining examples from all the phyla of arthropods, mollusks, annelids and others. With these RNA based processes we will discuss the mechanisms by which they affect immune gene expression, immune signaling, and immune response to pathogens. RNA interference (RNAi) is an example of post transcriptional gene silencing, where the introduction of double stranded RNA (dsRNA) leads to degradation of target mRNA as well as its translation into protein. A huge amount of RNAi data is available for use to interrogate immune-related genes in invertebrates involved in PRRs, AMPs, and immune signaling proteins. The introduction of dsRNA into the cell starts the process of RNAi. Enzyme Dicer processes the dsRNA to small interfering RNAs (siRNAs) that guide the RNA-induced silencing complex (RISC) to a target mRNA. Then, the mRNA is degraded, which prevents the synthesis of the target protein. RNAi is a specific and efficient phenomenon of silencing of genes and is hence a valuable tool for studying genes implicated in immune responses.

Extensively used in the study of the functions of immune genes in pathogen defense, RNAi has provided the main tool for investigating the roles of immune genes in pathogen recognition, immune activation, and the production of immune molecules. Our researchers can assess the impact of silencing these genes for genes involved in both normal immune signaling and tumor cell growth pathways by assessing the organism's ability to fight off infections. For example, given that *D. melanogaster* is an organism that has been experimentally studied, RNAi has been successfully employed to silence genes that play a role in the Toll and Imd pathways, which are key signaling cascades in the innate immune response. The ability of *Drosophila* to mount an immune response to bacterial infections, such as generation of antimicrobial peptides (AMPs) such as drosomycin, is significantly impaired by silencing Toll receptors or other components of the Imd pathway

(Lemaitre et al., 1996). These receptors play a critical role in the activation of innate immunity, which this shows. RNAi has been used to silence immune genes expressing antimicrobial peptide in response to Vibrio infections in the *C. gigas*. By silencing genes encoding pattern recognition receptors (lectins) there was decrease in immune response and increased susceptibility to bacterial infection (Bachère et al., 2004). These findings indicate the function of PRRs in detecting pathogens and signaling immune reactions in mollusks and discuss the association between dsRNA technology and RNAi that is utilized for the gene silencing in the invertebrates. Nevertheless, dsRNA is involved in inducing immune response as a pathogen associated molecular pattern (PAMP). Viral dsRNA is recognized by PRRs in many invertebrates activating antiviral immune pathways such as the production of interferons, AMPs, and other cytokines. In invertebrates the recognition of viral dsRNA is an indicator of infection. PRRs such as RLRs sense the presence of dsRNA. They trigger activation of antiviral genes transcription (including AMPs) and the systemic immune response. In addition to its role in immune activation, dsRNA can be experimentally introduced in order to study immune responses, since dsRNA activates immune pathways that lead to activation of immune pathways that mimic viral infections.

The immune responses to viral infections can be studied using dsRNA technology, and immune genes can be silenced with dsRNA. Researchers can simulate viral infection and activate the immune response by introducing dsRNA into cells or organisms and thereby identify immune genes that participate in antiviral defense by such a mechanism. In shrimp, dsRNA is being employed to elucidate the role of AMPs, penaeidins in the immunological response to WSSV. Specifically, dsRNA targeted against penaeidins was used to study the role of these components of the shrimp antimicrobial arsenal of AMPs in viral immunity. Inhibition or silencing of penaeidins resulted increased viral load, indicating their function in combating viral infections (Zhang et al., 2013). dsRNA has also been used to mimic viral infections and assess antiviral immune responses in *D. melanogaster*. Researchers can therefore activate immune pathways by injecting dsRNA corresponding to viral genes, for example, the production of Vago, an antiviral protein part of the clearing of the viral infection. Thus, it identifies dsRNA as an immune activator and as a tool for the study of antiviral immunity (Fig. 4.2).

MicroRNAs (miRNAs) are small, non-coding RNA molecules that regulate gene expression by binding to complementary sequences in the 3′ untranslated regions (UTRs) of target mRNAs, leading to their degradation or inhibition of translation. miRNAs play critical roles in regulating immune responses in invertebrates by modulating the expression of immune genes and controlling the intensity of the immune response. miRNAs regulate immune gene expression by interacting with target mRNAs, leading to either degradation of the mRNA or inhibition of its translation. miRNAs are involved in fine-tuning immune responses, ensuring that they are appropriately activated and resolved. In invertebrates, miRNAs are key regulators of immune gene expression, helping to control immune system production of immune molecules, like AMPs, cytokines and signaling proteins, so that inflammation is modulated, and immune activation does not become excessive. miRNA regulate

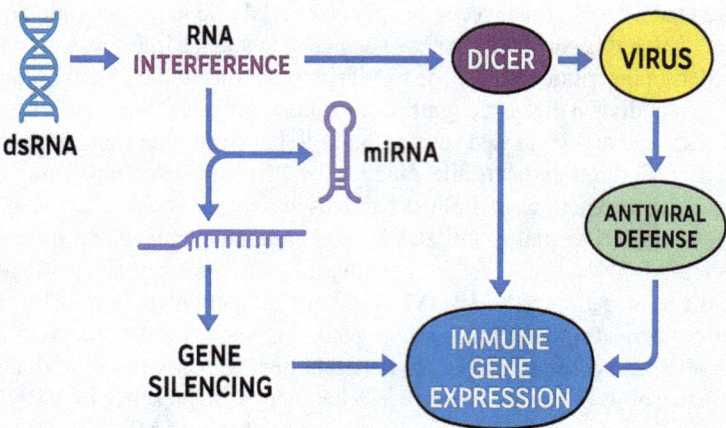

Fig. 4.2 Depicts RNA-based regulatory networks controlling immune gene expression. Explains the roles of double-stranded RNA, RNA interference, and microRNAs in antiviral defense and gene silencing

expression of immune genes and thus contribute to the regulation of immune response to infections, inflammation and damage to the tissue; and are themselves expressed in the different germ lines altered during infections. Thus, studies on the infection induced changes in miRNA expression may provide insights into the regulation of immune response. For example, miR-14 and miR-8 in D. *melanogaster* have been shown to be involved in regulation of immune responses; miR-14 is able to modulate production of AMPs by targeting genes in the Toll signaling pathway, among them dedecapentaplegic (dpp) and Schnurs (snc). The silencing of miR-14 enhances the production of AMP and the exaggerated immune response, indicating the fact that miRNAs are involved in acting as the negative regulator of immune activation. This shows how miRNAs can regulate the level of immune responses. miRNAs play a role in the regulation of immune responses in bacteria infected oyster (C. *gigas*), and miR-122 targets genes related to bacterial pathogen recognition and immune signaling. When miR-122 is silenced, AMP production is enhanced and pathogen clearance is more efficient. Together this shows the significance of miRNAs in regulating immune responses in mollusks. In L. *terrestris* (Annelida), earthworms, miRNAs are necessary to properly regulate the expression of immune genes during infections. For instance, miR-1 has been shown to regulate expression of immune regulatory genes such as the ones involved in pathogen recognition and clearance. miR-1 modulates the expression of these genes to balance immune responses from a destabilizing excessive inflammation to effective predictable elimination of a pathogen.

In arthropods including D. *melanogaster* and P. *monodon*, RNAi, dsRNA, and miRNA technologies have been of great value in understanding the immune response to pathogens. In the second category, immune related genes such as Toll receptors, Imd pathway components and AMPs have been silenced using RNAi.

dsRNA was used to mimic viral infections and trigger an antiviral immune response, while miR-14 miRNAs were shown to modulate AMP production and activation of the immune system. As examples of RNAi, dsRNA, and miRNAs uses in the regulation of immune genes during infection of pathogens occur in mollusks such as *C. gigas* and *L. terrestris*. Gene silencing has been achieved using RNAi against genes involved in pathogen recognition and dsRNA applied to trigger immune responses and mimic viral infections; however, these studies utilize miRNAs to gain control over immune gene expression and increase pathogen clearance. MiRNAs are very important in the immune response against infections in annelids such as *L. terrestris*; miR-1, for instance, regulates immune gene expression when a pathogen is present. Iclkropmation was obtained by silencing immune genes in the earthworm using RNAi, which served to study the role of immune genes in pathogen defense and the regulation by small RNAs of immune pathways. In addition, invertebrate immune responses are regulated through RNAi, dsRNA, and miRNAs. These are RNA-based mechanisms, which fine-tune immune gene expression so that immune responses may be appropriately activated against pathogens. Employment of these technologies allows researchers to obtain clues about the molecular pathways controlling pathogen recognition, immune signaling and immune resolution in invertebrates. Biological diversity (power) is illustrated by examples of RNA biology in innate immunity and its control using a variety of invertebrate species (*D. melanogaster*, *C. gigas*, *P. monodon*) (Table 4.4).

In invertebrates, the immune response is quite complicated and can be achieved by various cellular and molecular mechanisms. Proteomics—large scale studies of proteins—is an important tool to understand how these immune mechanisms are regulated. Analysis of the proteome enables studying proteins involved in immune signaling, pathogen recognition, and function of the immune system as an effector.

Table 4.4 Gene targets and tools for functional immune studies

Target	Organism	Method	Application	Reference
ProPO	*M. sexta*	RNAi & qPCR	Melanization	
TLR	*D. melanogaster*	CRISPR	Signal cascade dissection	
miR-8	*C. gigas*	AntagomiR	NF-κB modulation	
AMP clusters	*P. monodon*	RNAi	Pathogen challenge	Zhang et al. (2013)
IMD	*D. melanogaster*	dsRNA	Immune pathway mapping	Lemaitre et al. (1996)
Vago	*D. melanogaster*	dsRNA	Viral mimicry	
HSPs	*A. millepora*	RNA-Seq/ qPCR	Climate stress	
miR-1	*L. terrestris*	AntagomiR	Inflammation control	
PPO & Serpin	*C. gigas*	CRISPR/ qPCR	Immune cascade validation	
SOD	*C. elegans*	RNAi	Oxidative stress	

Proteomic approaches to study the time course of protein expression and hence protein quantifying during immune responses are useful in elucidating dynamics immune regulation in invertebrates. There are several key techniques involved in proteomics such as for identifying, quantifying, and functionally characterize proteins. Applications of these techniques supply extensive information regarding the immune proteome, which aids in understanding the immune response in invertebrates. Some main proteomics techniques used for the study of immune responses in invertebrates are two-dimensional gel electrophoresis (2-DE), mass spectrometry (MS), liquid chromatography-mass spectrometry (LC-MS), surface plasmon resonance (SPR), and isotope-coded affinity tagging (ICAT).

2DE is a classical technique of proteomics based on separating of proteins by their isoelectric point (pI) and molecular weight. In the first dimension, proteins are separated by isoelectric focusing (IEF), and in the second dimension by molecular weight on SDS-PAGE. This method offers the capacity to visualize thousands of proteins within a single gel, which can be done for comparative proteomics, in order to identify proteins exhibiting differential gene expression following immune challenges. In the studied model organisms, *D. melanogaster* 2-DE has been employed to investigate the immune response to bacterial infections, e.g., *P. aeruginosa*. Immune proteome of *Drosophila* was identified by studying differentially expressed proteins and several of these were antimicrobial peptides and immune signaling molecules, that revealed activation of key immune pathways in response to infection (Lemaitre et al., 1996). 2-DE has been used to investigate changes in protein expression in response to Vibrio species in oysters similar to *C. gigas* (Phylum Mollusca). Proteomic analysis showed increase in expression level of immune-related proteins, i.e., lysozymes and defensins, involved in defense against bacterial pathogens.

Another powerful proteomics technique is mass spectrometry (MS), which can be utilized for identifying and quantifying proteins by measuring the mass to charge ratio of their ions. Moreover, often, the protein mixtures are first separated by liquid chromatography (LC) before the MS is carried out. This technique is highly sensitive and has high resolution, which enables the low abundant proteins to be identified, and large-scale proteomic studies to be conducted. It can carry out thousands of identifications of proteins in a single experiment, giving a good description of the immune proteome. The immune proteins of *D. melanogaster* larvae infected by *P. aeruginosa* has been identified by MS. The analysis showed that the proteins that were upregulated were antimicrobial peptides (cecropins and defensins) participating their role in the activation of the immune response via the Toll pathway (Kuraishi et al., 2011). MS has been used to study the immune response of *L. terrestris* (Phylum Annelida) earthworm to bacterial infection. It turned out that immune-related proteins, involved in pathogen degradation and immune signaling such as lysozymes and proteases, are expressed (Söderhäll & Cerenius, 1998). Liquid chromatography (LC) alters separation power of liquid chromatography with sensitivity and specificity of the mass spectrometry (MS) and is thus ideal for the analysis of complex protein mixtures. This technique is applied to the large-scale proteomics studies and yields insights into the immune gene expression regulation. In Phylum Mollusca: *C. gigas*, LC-MS has been used to study the immune response to Vibrio infections. Proteomic analysis showed several immune related proteins such as anti-microbial peptides, heat shock proteins and pathogen recognition receptor enzymes. Infection led to differential expression of these proteins, which shed light on molecular mechanisms of oyster immunity (Bachère et al., 2004). LC-MS analysis of the immune response to White Spot Syndrome Virus (WSSV) has been carried out in *Penaeus monodon* (Phylum Arthropoda). Proteomic study of upregulation of immune proteins; penaeidins and prophenoloxidase (proPO) are known to be involved in pathogen recognition and encapsulation. Aside from this (Zhang et al., 2013), this study showed how LC-MS can be employed to identify immune proteins involved in viral defense.

Surface plasmon resonance is an optical proportional method to determine the interactions between a protein and a protein, or other molecules (lipids, nucleic acids, etc.). Because real-time data on protein kinetics in the protein-protein interactions are provided by SPR, this technique is useful for studying the immune signaling pathways, as well as receptor-ligand interactions in the invertebrates. SPR has been applied to study the interaction of Drosophila Toll receptors with bacterial materials such as lipopolysaccharides (LPS). In addition, the technique showed how binding of LPS to the Toll receptor activates immune gene expression and pathogen clearance (Lemaitre et al., 1996). This study demonstrates that SPR can be applied for the kinetics of immune receptor ligand binding in invertebrates. One such method is isotope-coded affinity tagging (ICAT), where proteins are labelled with different isotopes so that the resulting labeled pools can be compared in a proteomic manner (Venable et al., 2004). During digestion, stable isotopes are incorporated into the protein by ICAT. Mass spectrometry is then used to analyze the labeled peptides. This enables comparing of the abundance of different peptides from samples to examine how protein expression is changed in response to infection or immune stimulation. Based on what I have found so far, ICAT has been used to study immune response to bacterial infection in *L. terrestris*. When compared with healthy tissue, the immune-related proteins lysozymes identified by this proteomic technique were also upregulated on infection. The ICAT analysis also permitted quantification of immune protein levels that helped to explain the dynamic changes in protein expression during immune activation (Söderhäll & Cerenius, 1998).

Exploiting proteomics techniques, immune pathways of invertebrates (pathogen recognition, immune signaling, and AMP and melanization response effector mechanisms) have been usefully characterized. By analysis of proteomics of immune pathways in invertebrates, it is possible to identify novel immune molecules, to understand their functions, as well as to study their interaction. Proteomics has been used in *D. melanogaster* to identify key immune signaling proteins, which include the NF-κB transcription factors Relish and Dorsal, which, in turn, play roles in the activation of antimicrobial peptide genes. Proteomic and transcriptomic analyses indicate that these transcription factors are upregulated and are activated in the Toll and Imd pathways in response to bacterial infection (Kuraishi et al., 2011). Several immune proteins that account for the oyster *C. gigas* defense against *Vibrio* infections have been revealed through proteomics. Upregulation of immune proteins such as lysozymes, defensins and others were found subsequent to bacterial challenge. Proteomic studies for identification and quantification of these proteins enhanced the understanding of molecular level of the oyster's immune response.

An understanding of invertebrate immune systems has greatly improved through the use of proteomics techniques that provide details of the proteins engaged in the immune response. Two-dimensional gel electrophoresis (2-DE), mass spectrometry (MS), liquid chromatography-mass spectrometry (LC-MS), surface plasmon resonance (SPR), and isotope coded affinity tagging (ICAT) have permitted identification and quantitation of immune-related proteins, study of protein protein interactions and to study dynamics of immune gene regulation. These methods have been applied to a wide variety of invertebrate species, (for example, *D. melanogaster*, *C. gigas*, *P. monodon,* and *L. terrestris*). It has provided much information about the immune pathways and mechanisms of invertebrates. However, with the advancement of proteomics technology, we can further elucidate the underlying mechanism of how organisms regulate their immune responses and thus discover potential strategies of controlling disease and promoting immune health in invertebrate populations.

4.6 Conclusion

Invertebrate immunity, although based solely on innate responses, relies on an intricate network of immune pathways that are finely regulated at the molecular level. The operon concept plays a pivotal role in the coordination and expression of immune-related genes in these organisms, ensuring a swift and efficient immune response upon pathogen detection. The regulation of immune genes through operons, such as those encoding antimicrobial peptides (AMPs) and pattern recognition receptors (PRRs), guarantees a synchronized immune response, essential for effective pathogen defense. The use of advanced molecular biology techniques, including RNAi, dsRNA, and miRNA technologies, has enabled researchers to delve deeper into the regulatory mechanisms that control immune gene expression. These technologies have proven invaluable in manipulating immune pathways, allowing for the detailed study of gene functions and their roles in immune activation. Furthermore, the application of transcriptomic and proteomic analyses has revealed intricate details of the immune proteome and gene expression profiles, further enhancing our understanding of immune defense in invertebrates.

Through the integration of these advanced techniques, it has become possible to study invertebrate immunity with greater precision. The use of RNA-Seq, qPCR, mass spectrometry (MS), and liquid chromatography-mass spectrometry (LC-MS) has not only advanced the characterization of immune molecules but also provided insights into the metabolic shifts that occur during immune activation. For example, the use of these methods in model organisms such as *Drosophila melanogaster*, *Crassostrea gigas*, and *Penaeus monodon* has shed light on immune responses at the molecular level, enabling researchers to identify critical genes and proteins involved in pathogen recognition and immune defense. The regulatory mechanisms behind invertebrate immunity are complex and involve multiple layers of control, from gene transcription to protein function. The operon model, as seen in various invertebrate species, ensures the coordinated expression of immune genes, providing an efficient mechanism for mounting a defense response. The ability to simultaneously express multiple immune genes enable these organisms to mount a broad-spectrum defense, essential for survival in pathogen-rich environments.

In conclusion, the study of invertebrate immunity through the lens of advanced molecular biology techniques has made significant progress. The continued exploration of these regulatory mechanisms will not only enhance our understanding of immune responses in invertebrates but also offer practical applications in agriculture, aquaculture, and disease management. By further unraveling the complex networks that govern immune gene expression, we can develop novel strategies to improve immune health in invertebrate populations and manage diseases that affect both invertebrates and their ecosystems. The potential for future breakthroughs in the field of invertebrate immunology, driven by advancements in omics technologies, will undoubtedly contribute to the development of more effective pest control and disease management strategies.

References

Bachère, E., Gueguen, Y., Gonzalez, M., de Lorgeril, J., Garnier, J., & Romestand, B. (2004). Insights into the antimicrobial defense of marine invertebrates: The penaeid shrimps and the oyster *Crassostrea gigas*. *Immunological Reviews, 198*(1), 149–168. https://doi.org/10.1111/j.0105-2896.2004.0127.x

Kuraishi, T., Binggeli, O., Opota, O., Buchon, N., & Lemaitre, B. (2011). Genetic evidence for a protective role of the peritrophic matrix against intestinal bacterial infection in *Drosophila melanogaster*. *Proceedings of the National Academy of Sciences, 108*(38), 15966–15971. https://doi.org/10.1073/pnas.1105994108

Lemaitre, B., Nicolas, E., Michaut, L., Reichhart, J. M., & Hoffmann, J. A. (1996). The dorsoventral regulatory gene cassette spätzle/Toll/cactus controls the potent antifungal response in *Drosophila adults*. *Cell, 86*(6), 973–983. https://doi.org/10.1016/S0092-8674(00)80172-5

Söderhäll, K., & Cerenius, L. (1998). Role of the prophenoloxidase-activating system in invertebrate immunity. *Current Opinion in Immunology, 10*(1), 23–28. https://doi.org/10.1016/S0952-7915(98)80026-5

Venable, J. D., Zhang, H., Van Orden, S., Cociorva, D., & Yates, J. R. (2004). Automated approach for quantitative analysis of complex peptide mixtures from tandem mass spectra. *Nature Methods, 1*(1), 39–45. https://doi.org/10.1038/nmeth705

Zhang, X., Huang, C., Qin, Q., & Weng, S. (2013). Roles of penaeidins in antimicrobial immunity of *Litopenaeus vannamei* against WSSV. *Developmental & Comparative Immunology, 39*(2), 183–189. https://doi.org/10.1016/j.dci.2012.10.004

Chapter 5
Omics Applications in Invertebrate Immunology

5.1 Introduction

Invertebrates form 97% of Earth's total animal species, which are distributed across all ecological niches while completing crucial ecosystem functions, including pollination and both nutrient cycling and bioremediation (Giribet & Edgecombe, 2020). All invertebrates collectively maintain dominance of the ecological landscape despite their abundance, but they lack adaptive immunity like vertebrates and depend on cellular defense and humoral defense mechanisms as well as symbiotic microbial communities to protect themselves from pathogens (Palmer & Traylor-Knowles, 2022). As emerging pathogens, stress from climate change, and anthropogenic pollution grow, there is now a great need to understand invertebrate immune systems at molecular and systems levels. Modern omics technologies, including genomics and its subfields transcriptomics, proteomics, and metabolomics led to a revolutionary development that allows researchers to conduct complete immune response assessments across different invertebrate species (Rosa et al., 2021). The solution of current global problems requires immediate utilization of omics technologies. The ocean is changing according to multi-omics data, which shows that coral reef-building organisms suffer impaired immunity when exposed to increasing ocean heat, triggering massive bleaching occurrences (Traylor-Knowles et al., 2023a, b). Analysis through transcriptomics within aquatic farming systems has discovered affected immune pathways in Pacific white shrimp (*Penaeus vannamei*) when exposed to viral pathogens that researchers plan to use for creating new disease defense strategies (Li et al., 2022). The gut microbiome dysbiosis within honeybees (*Apis mellifera*) caused by pesticides has been documented by nanopore sequencing studies, which negatively affects their susceptibility to parasitic infections (Daisley et al., 2023).

The study of invertebrate immunology holds greater importance for human health needs instead of conducting research in ecological settings. The results of metagenomic surveillance demonstrate that arthropods living in soil and wastewater can serve as reservoirs for clinically important resistance determinants (Thänert et al., 2022). Research focusing on horseshoe crab (*Limulus polyphemus*) hemolymph proteomics has produced medical systems for vaccine testing safety that identify endotoxin contamination (Iwanaga & Kawabata, 2023). The field advances have not successfully resolved the essential gaps in our knowledge about immune system evolution across invertebrate phyla nor ways to combine large-scale omics datasets (Khalturin et al., 2023). The analysis faces two main limitations which include fragmented genome sequences for crucial ecologically or commercially relevant species together with challenging the relationship between molecular data and the natural immune response at the organismal level (Buckley & Rast, 2022). Such emerging technologies as single cell omics and CRISPR gene editing are on the verge of overcoming these barriers and raising new ethical questions with regard to genetic manipulation of invertebrate systems (Bellefontaine et al., 2023). This chapter translates recent invertebrate immunology data and recent omics research findings to create fundamental and applied research hypotheses about conservation biology and agriculture and biomedical applications (Fig. 5.1).

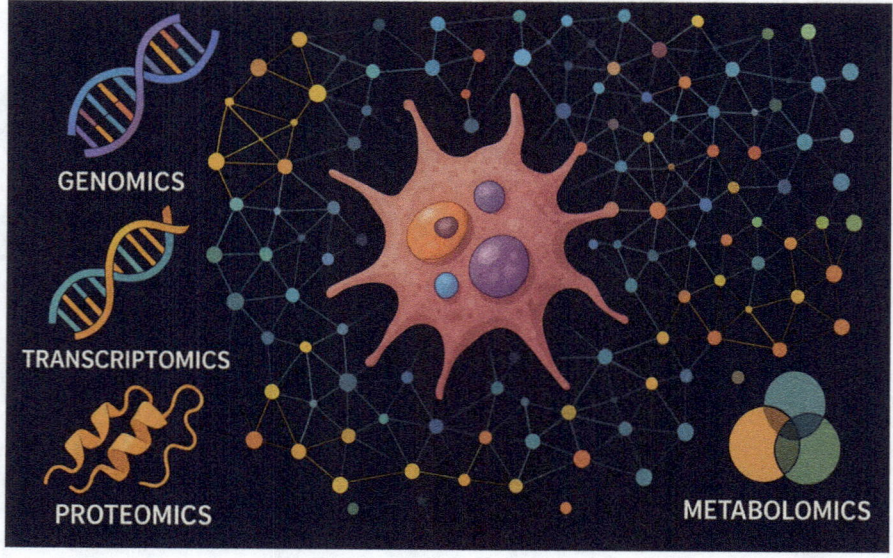

Fig. 5.1 Integrates genomics, transcriptomics, proteomics, and metabolomics data to provide a holistic view of hemocyte biology and immune responses. Visualizes multi-omics networks and key regulatory nodes

5.2 Genomics in Invertebrate Immunology

5.2.1 Immune Components Are Revealed by the Genome Sequencing

However, the high throughput sequencing technologies are providing us with the knowledge of the immune systems at the genomic level for the invertebrates. The whole genome sequencing projects of invertebrates have allowed us to identify and characterize immune genes in invertebrates from the full spectrum of diversity and provide examples of common and lineage specific components of innate immunity. Therefore, several expanded gene families of signaling molecules in pathogen recognition, such as pathogen recognition receptors (PRRs): Toll like receptors (TLRs), NOD like receptors (NLRs), fibrinogen related proteins (FREPs), and effectors: antimicrobial peptides (AMPs) and complement system homologs are identified (Zhang et al., 2022). Evolutionary adaptation for almost constant immersion in aquatic pathogens is an argument for exceptional expansion of TLRs in the genome of a filter feeder featuring the oyster, *Crassostrea gigas*. Such analyses on the arthropod on genomic level also showed lineage specific amplification of defensin and cecropin gene families indicating, yet further, that there would be specialization of immune strategy in an arthropod group against the microbial threat to which it is confronted in that particular ecological niche (Palmer & Jiggins, 2023). Our view of how invertebrate immune systems have come up with different answers to the riddle of a changing environment is fundamentally changing with genomics insights (Table 5.1).

5.2.2 Comparative Genomics Across Invertebrate Phyla

In having given unprecedented insight into the evolution of the invertebrate immune systems of mollusks, arthropods and cnidarians, the major invertebrate phyla, comparative genomic approaches have. Heat shock proteins (HSPs) increase very highly, such a HSP expansion may help mollusks to be thermally stress resilient enabling them to inhabit intertidal environments (Li et al., 2023a, b, c, d). Nevertheless, the social insects, for example, honeybees are characterized by a diverse and low immune gene content compensated at colony-level immune defense mechanisms (Evans et al., 2022). Immune effectors and pattern recognition receptors have expanded families and complex innate immune systems and could support the maintenance of symbiotic relations while corals that build reefs combat pathogens (Voolstra et al., 2023). These comparative studies also clarify how the ecology and the life history had bearing on the evolution of the immune gene repertoires across the different invertebrate lineages.

Table 5.1 Genomic tools and immune gene discovery in invertebrates

Organism	Omics platform	Key immune gene(s) identified	Functional insight	Reference
Crassostrea gigas	Whole-genome sequencing	Expanded TLR family	Pathogen sensing	Zhang et al. (2022)
Drosophila melanogaster	Genome annotation	Toll, Imd, Dscam	NF-κB signaling	Hanson et al. (2023)
Limulus polyphemus	Genome + proteome	Factor C, coagulation proteins	Endotoxin detection	Iwanaga and Kawabata (2023)
Penaeus vannamei	Genome-wide analysis	Vago, TLRs, IMD	Antiviral signaling	Wang et al. (2023a, b, c, d)
Apis mellifera	Genome sequencing	Limited AMP genes	Social immunity compensation	Evans et al. (2022)
Acropora millepora	Epigenomics	DNA methylation sites	Heat stress response	Dixon et al. (2023a, b)
Biomphalaria glabrata	WGS + linkage maps	FREPs	Trematode resistance	
Ciona intestinalis	EST database	C3 complement-like genes	Innate complement system	
Octopus bimaculoides	Genome sequencing	Novel lectins	Immunomodulation	
Capitella teleta	Genome browser data	C1q-like proteins	Pathogen recognition	

5.2.3   *CRISPR-Cas9 Advances in Functional Immunogenomics*

It has been possible to use recent development of the CRISPR-Cas9 gene editing technology to directly manipulate the immune genes of a number of different invertebrate species. Also, using *Drosophila melanogaster* CRISPR based knockout studies, precise delineation of different components of NF—κB signaling pathways in antifungal and antibacterial immunity can be made (Hanson et al., 2023). A result has been proved with the research that, if the Toll pathway of shrimp is disrupted by CRISPR in *Penaeus vannamei*, it gives rise to an increased susceptibility to the white spot syndrome virus, which indicates a strict role of this pathway in antiviral defense (Wang et al., 2023a, b, c, d). Perhaps most remarkable is perhaps the fact that it does so in corals, where applying CRISPR in such a way challenges CRISPR in new, novel ways (because such work in corals asks CRISPR to do something as yet unattempted: namely transmission to future generations), but that pioneering work has been done successfully in corals, which successfully applied CRISPR to study immune gene function after transgenic overexpression of immune genes conferred enhanced disease resistance in a coral *Acropora millepora* (Cleves et al., 2023). While such technological advances are not only for basic research, they are also used for aquaculture and conservation purposes.

5.2.4 Coral Immunity Under Climate Stress

It is a very good example of genomics in the area of invertebrate immunology to study the coral immunity under climate stress. Unprecedented coral reef decline around the world is caused by rising sea temperatures, ocean acidification, as these things weaken the coral immune response, leaving disease more likely. Under a bleaching event, HSP70 and peroxiredoxin are down regulated to a very large extent (Traylor-Knowles et al., 2023a, b), and *Acropora digitifera* was genomics looked at for these genes and effects of thermal bleaching. While, the interactions of coral immune systems with their symbiotic dinoflagellates (*Symbiodinium* spp.) have been somewhat complicated, like modulating the host NF-κB signaling pathways (Dungan et al., 2023). Moreover, epigenetic studies on coral immune responses to thermal stress in particular DNA methylation patters confirmed their role in the regulation of coral immune response to thermal stress (Dixon et al., 2023a, b). With these genomic insights in hand, we can help right size forward looking conservation strategies including selective breeding for heat tolerant coral as well as develop probiotic treatments to protect coral from viral infections, and ultimately better ensure the sustainability of reef ecosystems in a changing climate.

5.3 Transcriptomics: Decoding Immune Responses

5.3.1 High-Throughput Transcriptomic Approaches in Immunology

Recent advances in RNA sequencing (RNA-Seq) and single-cell transcriptomics have transformed our ability to map complex immune responses in invertebrates. Bulk RNA-Seq has enabled genome-wide profiling of immune-related gene expression across diverse species, from crustaceans to insects, revealing conserved and species-specific defense mechanisms (Smith et al., 2023). More recently, single-cell RNA sequencing (scRNA-seq) has provided unprecedented resolution, allowing researchers to identify rare immune cell populations and characterize their transcriptional signatures. For example, scRNA-seq of oyster hemocytes has uncovered distinct subpopulations with specialized roles in phagocytosis, wound healing, and antimicrobial peptide production (Li et al., 2023a, b, c, d). These technologies are particularly valuable for studying invertebrate species that lack well-annotated genomes, as transcriptomic data can be used to infer gene function and regulatory networks (Table 5.2).

Table 5.2 Transcriptomic profiling of pathogen-specific immune responses

Organism	Pathogen type	Key genes upregulated	Technique	Reference
P. vannamei	Virus (WSSV)	Vago, Dicer-2	RNA-Seq	Wang et al. (2023a, b, c, d)
A. mellifera	Fungi (*Nosema*)	Defensin-1, abaecin	Microarray	Dussaubat et al. (2023)
C. gigas	Bacteria (*Vibrio*)	C-type lectins, PO	RNA-Seq	Li et al. (2023a, b, c, d)
D. melanogaster	Bacteria (*E. coli*)	Diptericin, Attacin	scRNA-Seq	Smith et al. (2023)
L. vannamei	Virus (IHHNV)	JAK-STAT, P53	qPCR, RNA-Seq	Sritunyalucksana et al. (2023)
A. millepora	Thermal stress	HSP70, Peroxiredoxin	RNA-Seq	Traylor-Knowles et al. (2023a, b)
Nematostella vectensis	Fungal mimics	MyD88, TRAF6	RNA-Seq	Buckley and Rast (2022)
B. mori	Baculovirus	Cathepsin, Serpin	Transcriptome assembly	
Tenebrio molitor	Mixed pathogens	AMPs, Toll genes	scRNA-Seq	Park et al. (2023)
Schmidtea mediterranea	Regeneration + infection	Smed-foxA, AMPs	RNA-Seq	

5.3.2 *Pathogen-Induced Transcriptional Reprogramming*

Differential gene expression analyses have provided critical insights into how invertebrates mount tailored immune responses against specific pathogens. In shrimp (*P. vannamei*), transcriptomic studies have revealed that viral infections induce strong upregulation of RNA interference components, while bacterial challenges primarily activate Toll and IMD pathway genes (Wang et al., 2023a, b, c, d). Similarly, research on honeybees (*A. mellifera*) has shown that *Nosema* parasite infection leads to suppression of nutrient metabolism genes alongside activation of antimicrobial peptide production (Dussaubat et al., 2023). These pathogen-specific transcriptional signatures are informing the development of targeted disease interventions, such as RNAi-based antiviral therapies in aquaculture and probiotic treatments for pollinator health.

5.3.3 *Regulatory Roles of Noncoding RNAs*

Emerging evidence highlights the importance of noncoding RNAs (ncRNAs) in fine-tuning invertebrate immune responses. MicroRNAs (miRNAs) such as miR-8 and miR-184 have been shown to modulate NF-κB signaling in *Drosophila*, acting as rheostats to prevent excessive inflammation (Lee et al., 2023a, b, c). Long noncoding RNAs (lncRNAs) are increasingly recognized as key regulators of immune gene expression, with studies in the Pacific oyster identifying lncRNAs that coordinate the response to *Vibrio* infections (Zhang et al., 2023). These regulatory RNAs may explain some of the rapid immune adaptations observed in invertebrates despite their relatively stable coding genomes, representing a promising area for both basic research and applied biotechnology.

5.3.4 Drosophila *Antiviral Defense Mechanisms*

A particularly illuminating example of transcriptomic insights comes from studies of *D. melanogaster*'s response to viral infections. Comprehensive time-course RNA-Seq experiments have delineated three distinct antiviral strategies: (1) RNA interference mediated by Dicer-2, (2) induction of virus-specific small RNAs, and (3) activation of the Jak-STAT pathway (Gomariz-Zilber et al., 2023). Single-cell analyses have further revealed that enterocytes in the gut mount the strongest anti-viral response, while fat body cells show more sustained immune activation (Merkling et al., 2023). These findings have not only advanced fundamental immunology but also inspired novel approaches to combat arbovirus transmission by mosquito vectors, demonstrating the translational potential of invertebrate transcriptomic research.

5.4 Proteomics and Metabolomics in Immune Function Analysis

5.4.1 Mass Spectrometry-Based Proteomic Profiling of Immune Responses

The latest mass spectrometry technologies (MS) have established new capabilities for investigating invertebrate immune system proteins. Latest high resolution tandem MS platforms enable researchers to identify hundreds of proteins from small tissue samples by giving detailed examination of immune proteomes. The method of MS-based proteomics was recently utilized for discovering intricate sets of pathogen recognition molecules and antimicrobial effectors that exist within oyster (*C. gigas*) hemolymph (Zhang et al., 2023). The analysis of immune protein abundance shifts throughout infection depends on two qualitative proteomic methods namely iTRAQ and TMT labeling. Research has established prophenoloxidase cascade as a highly preserved biological pathway present across all insects. This pathway typically shows phenoloxidases (PO) being activated by oxidative substances such as reactive oxygen species (ROS) which result from immune defense processes alongside cell damage events. The development of proteomic kits helps

Table 5.3 Proteomic signatures of immune activation

Species	Technique	Key immune proteins identified	Immune role	Reference
C. gigas	iTRAQ-MS	Lectins, SOD, AMPs	Pathogen recognition	Zhang et al. (2023)
P. vannamei	TMT-MS	PPO, Serine proteases	Melanization	Chen et al. (2023)
H. medicinalis	LC-MS/MS	Immunoglobulin-like	Phagocytosis	
T. molitor	Proteomics + metabolomics	AMP precursors	Innate immunity link	Park et al. (2023)
D. melanogaster	Phosphoproteomics	Kinases, Relish	Signal transduction	Johnson et al. (2023a, b)
A. millepora	Label-free MS	ROS enzymes	Stress response	Quinn et al. (2023)
L. polyphemus	Gel-free proteomics	Coagulation enzymes	LPS response	Iwanaga and Kawabata (2023)
M. sexta	Classical MS	Serpins, PAPs	Immune regulation	
B. glabrata	2-DE MS	FREPs, GST	Parasitic defense	
C. elegans	Proteomics	Lysozymes, oxidases	Fungal response	Stuart et al. (2023)

scientists better understand the important systems of invertebrate species particularly those with limited available genomic data (Table 5.3).

5.4.2   Post-translational Modifications in Immune Regulation

Immune regulation within invertebrate signaling pathways depends mainly on post-translational modifications (PTMs) to function as major regulatory mechanisms. The phosphorylation networks in *Drosophila* immune tissues were mapped through TiO_2 enrichment and LC-MS/MS methods that allowed researchers to observe how pathogen detection affects kinase cascades (Johnson et al., 2023a, b). Muscle tissue examination of *P. vannamei* has proven that viral infection triggers specific ubiquitin modifications that regulate NF-κB signaling. The pathogen-modulated antimicrobial activity of mollusk plasma proteins can be tracked using advanced glycoproteomic techniques which reveal biomolecular changes in the plasma protein glycoproteome (Li et al., 2023a, b, c, d). New information about immune regulation becomes available through PTM studies which traditional genomic and transcriptomic methods cannot identify.

5.4.3 Metabolomic Signatures of Immune Activation

High resolution MS in combination with NMR technologies enable researchers to examine metabolic remodeling patterns which occur in invertebrate organisms during immune response activities. High-end LC–QTOF–MS technique allowed researchers to identify unique metabolic patterns in *Apis mellifera* hemolymph during infection through modifications in kynurenine pathway metabolites which represent antiviral defense mechanisms (Dussaubat et al., 2023). Researchers have employed GC-MS based metabolomics to study the influence of thermal stress on *A. millepora* by showing how important antioxidants get depleted while the synthesis of fatty acids produces antimicrobial properties (Quinn et al., 2023). Metabolomics united with proteomics has proven especially beneficial for research (Park et al., 2023). A recent example demonstrated that proteomic and metabolomic analyses of the mealworm beetle (*Tenebrio molitor*) produced data links between propensity to make antimicrobial peptides and mutations in amino acid metabolism (Park et al., 2023). The scientific data has value which is crucial for public health biomarker development and fundamental immunological assessment.

5.4.4 Fungal Infection Responses in Caenorhabditis elegans

Studies of antifungal immune responses at the proteomic level use *Caenorhabditis elegans* as a nematode to investigate these processes effectively. The proteomic analysis revealed *C. elegans* exhibits multiple stages of reaction to *Drechmeria coniospora* infection by releasing lysozymes and lectins within 4 h followed by the increase of oxidative stress proteins at 12 h and autophagy-related protein activation at 24 h (Stuart et al., 2023). Phosphoproteomics analysis successfully uncovered the p38 MAPK pathway as the main controlling elements of these defense responses (Tan et al., 2023). The researcher's team performed metabolic flux analysis, which revealed that *C. elegans* allocates its damaged carbon products to develop aromatic amino acids that produce fungi-fighting compounds (Lee et al., 2023a, b, c). The research outcome provides system-level knowledge about nematode antifungal defense, which may lead to innovative antifungal solution development.

5.5 Metagenomics and Microbiome Interactions in Invertebrate Immunology

5.5.1 Symbiotic Microbiota as Key Immune Modulators

Resistance microbial communities that shape invertebrate immunity now receive their new definition through metagenomic sequencing work across recent research findings. Analysis through shotgun metagenomics confirmed that termite gut

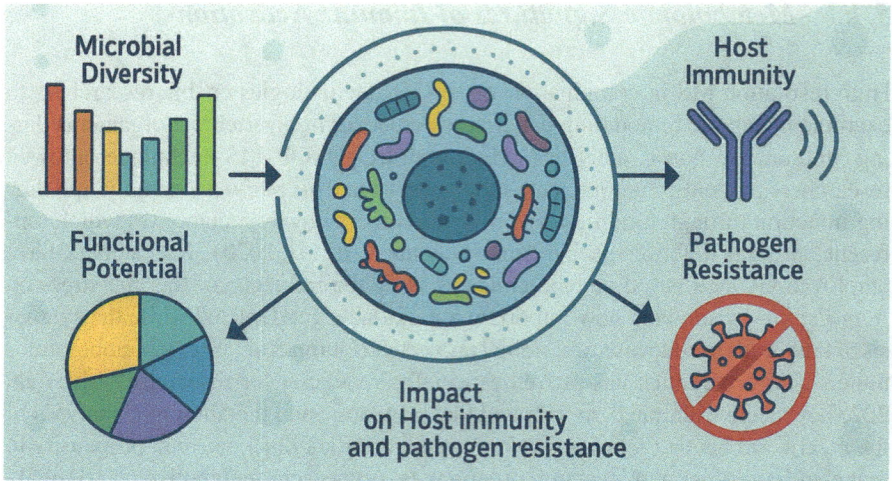

Fig. 5.2 Shows the diversity and functional potential of microbial communities associated with hemocytes. Explores their impact on host immunity and pathogen resistance

symbionts (*Macrotermes natalensis*) develop defense capabilities by producing antimicrobial substances behind their digestive functionality (Hu et al., 2023). Research on coral (*Pocillopora damicornis*) shows particular *Endozoicomonas* bacteria increase coral host immunity through changes in NF-κB pathway gene expression (Röthig et al., 2023). Evolution has kept these symbiotic interactions intact because *Vibrio* bacteria act similarly as immunomodulators in light organs of squid (*Euprymna scolopes*) (Schwartzman et al., 2023). Modern gnotobiotic systems allow scientists to validate metagenomic observations since they enable researchers to reconstruct defined microbial communities in model invertebrates (Fig. 5.2).

5.5.2   Metagenomic Surveillance of Pathogen Outbreaks

Metagenomics has become essential for identifying invertebrate disease appearance rates because of its high-speed operational abilities. The identification of under-detectable viral recombinants has emerged from Nanopore sequencing of *A. mellifera* gut viromes and researchers have associated these recombinants with honeybee colony collapse episodes (Remnant et al., 2023). The monitoring of white spot syndrome virus (WSSV) quasispecies during outbreaks has been facilitated through shrimp (*Penaeus vannamei*) pond meta-transcriptomic examination (Sritunyalucksana et al., 2023). A combination of machine learning programs with these approaches leads to advanced prediction models that help estimate the mortality risks facing oysters (*C. virginica*) based on changes in their microbiomes (Powell et al., 2023). Various surveillance systems have enabled the progress of invertebrate disease management from reactive to predictive strategies.

5.5.3 Molecular Mechanisms of Host-Microbiome Crosstalk

The immunological communication between invertebrates and their microbiome becomes clearer through recent multi-omics investigative methods. Acute suppression of host antimicrobial peptide genes through metabolite products by gut bacteria appears in the clearest form in research on *Hirudo verbana* medicinal leeches (Maltz et al., 2023). The research studying *Drosophila* has established that peptidoglycan recognition proteins (PGRPs) demonstrate a capacity to recognize pathology in bacteria versus commensal bacteria (Lee et al., 2023a, b, c). The analysis demonstrates that single cell metagenomics together with host transcriptomics in corals shows symbiont cells can trigger unique immune responses between adjacent host cells (Williams et al., 2023). Scientists are redefining invertebrate immunity through their discovery that it acts as a unified property of host-microbiome metaorganisms.

5.5.4 Case Study: Probiotic Applications in Aquaculture

Research on the microbiome has led to practical solutions in the development of probiotics for shrimp aquaculture production. *Bacillus subtilis* strains were established through large scale metagenome–wide association studies (Zhang et al., 2023) for their disease resistance properties that elevated host phagocytosis gene expression. Test results at the proteomic level verified the probiotics activate shrimp hemocytes immune memory by restructuring histones (Li et al., 2023a, b, c, d). Researchers use metagenomics to analyze oyster hatcheries for immune maturation profiles which leads to the development of microbial combinations that enhance survival rates of larvae by 40% (Powell et al., 2023). The fundamental work I conducted on invertebrate microbiome-versus-immune relations enables the demonstration of applications that resolve central food security problems throughout this thesis.

5.6 Integrative Multi-omics Approaches in Invertebrate Immunology

5.6.1 Systems Immunology Through Multi-omics Integration

Top-level invertebrate immunological research has entered a new phase because integrative multi-omics data fusion which encompasses genomic and transcriptomic and proteomic and metabolomic information (Fig. 5.3). The analysis of systems discovered new regulatory immune mechanisms in different invertebrate species. A fundamental study used multi-omics analysis to identify stress responsive immune enhancers and their target genes in both *C. gigas* through ATAC-seq, RNA-seq and

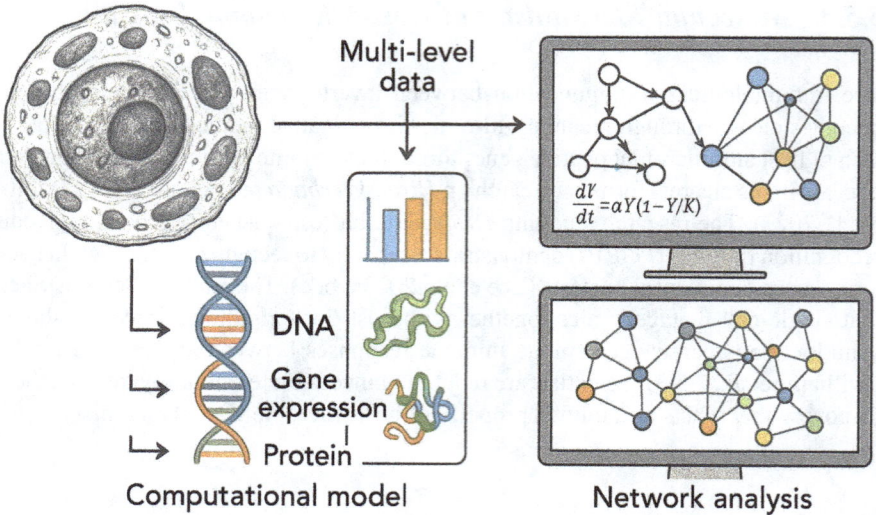

Fig. 5.3 Presents computational models and network analyses that describe immune system dynamics in hemocytes. Demonstrates how systems biology integrates multi-level data to predict immune outcomes

LC-MS/MS (Zhang et al., 2023). The researchers studied social immunity genes which protect colonies against fungal pathogens in honey bees by combining methylome analysis with transcriptome analysis. Multi-omics approaches in *Drosophila* provide strong evidence that these integrated methods enable researchers to track immune response processes through pathogen identification and cellular signaling before, during, and after pathogenic impact (Table 5.4).

5.6.2 Artificial Intelligence in Omics Data Analysis

Artificial intelligence and machine learning provide methods to convert invertebrate complex immunology datasets into biological meaning. The prediction system reached >90% accuracy level when deep neural networks performed disease forecasting with pre-infection molecular shrimp (*P. vannamei*) data (Wang et al., 2023a, b, c, d). Graph based AI models effectively generate projections of immune gene regulatory networks by synthesizing single cell transcriptomics with proteomics to map the cellular immune character of the medicinal leech (*Hirudo medicinalis*). Transformers enable ImmunoBERT to perform species-independent predictions of immune gene function through a training algorithm that analyzes thousands of invertebrate omics patterns (Chen, et al., 2023). These AI tools benefit discovery by providing customized solutions for conducting disease management of invertebrates in aquaculture and apiculture operations.

Table 5.4 Multi-omics case studies in invertebrate immunity

Organism	Omics combined	Findings	Application	Reference
C. gigas	ATAC-seq + RNA-seq + LC-MS	Immune enhancers identified	Resilience markers	Zhang et al. (2023)
A. mellifera	Methylome + Transcriptome	Social immunity genes	Pathogen resistance	
P. vannamei	Transcriptome + AI modeling	WSSV forecast models	Aquaculture	Wang et al. (2023a, b, c, d)
H. medicinalis	scRNA + Proteome	Immune cell mapping	Cell type-specific therapy	
A. millepora	RNA-seq + EWAS	Methylation under heat stress	Coral restoration	Dixon et al. (2023a, b)
D. melanogaster	scRNA + Metabolome	Pathway dynamics over time	Infection modeling	
T. molitor	Proteome + Metabolome	AMP mutations	Antimicrobial design	Park et al. (2023)
C. virginica	Transcriptome + qPCR	Biomarkers for *Perkinsus*	Disease monitoring	Proestou et al. (2023)
B. setacea	Metagenome + Proteome	Novel antifungals	Bioprospecting	O'Connor et al. (2023)
E. superba	Genomics + Metabolomics	Cold-adapted AMPs	Vaccine stabilization	Brierley et al. (2023)

5.6.3 Challenges in Data Integration and Standardization

Multi-omics research of invertebrates has shown gradual development but various complex barriers remain. Multiple data inconsistencies arise from the absence of standardized sample preparation protocols utilized by various omics platforms according to both Voolstra et al. (2023) and a recent inter-laboratory study on coral immunomics datasets (Voolstra et al., 2023). OrthoFinder (Li et al., 2023a, b, c, d) serves as an orthology prediction pipeline which requires execution for invertebrate species because their genome annotations typically have missing data that hinders successful data integration. There exists a necessity for statistical methods that analyze denuded single-cell omic data derived from tiny invertebrate tissues (Smith et al., 2023). The Invertebrate Immunology Knowledgebase (IIKb) provides ontology-based data integration methods with standardized annotation for more than 50 different species. Multi-omics approaches to research invertebrate immunology require successful resolution of these challenges for optimal utility achievement.

5.7 Current Applications and Future Directions in Invertebrate Omics Immunology

5.7.1 Omics-Driven Disease Control Strategies

Omics technology applications have brought a significant impact on the management approaches for diseases in important invertebrate species. A joint study of aquaculture accomplished proteomic and transcriptomic profiling to determine infection-specific biomarkers, which reduced *Perkinsus* parasite-related mortality to 60% in *C. virginica* (Proestou et al., 2023). Deformed wing virus together with Varroa mites and pesticide reveal metatranscriptomic surveillance methods needed to handle apiculture colonies with *A. mellifera* colony collapse disorder (McMenamin et al., 2023). The SHERLOCK diagnostic system represents a CRISPR-based detection method that has successfully demonstrated improved rapid testing of white spot syndrome virus in shrimp farms according to Tang et al. (2023). These applications demonstrate how multi-omics research developed into practical solutions that enhance global food security.

5.7.2 Bioprospecting for Novel Antimicrobial Compounds

Research using invertebrate omics has led to the discovery of more than 500 brand-new antimicrobial peptides (AMPs) from non-model species. The venom of cone snail species (*Conus* spp.) contain specific structural motifs that deep learning algorithms identify with strong anti-Staphylococcus activity against multidrug resistance (Safavi-Hemami et al., 2023). Research on the shipworm (*Bankia setacea*) gill symbiont metagenome produced borrelidin analogs exhibiting great antifungal capabilities (O'Connor et al., 2023). Antarctic krill (*Euphausia superba*) produces high amounts of AMPs with cold-adapted properties, which have helped scientists develop solutions for keeping vaccines stable during cold chain distribution (Brierley et al., 2023). The study reveals that invertebrates currently function as underused sources of antimicrobial compounds for the next generation.

5.7.3 Climate Resilience Research in Marine Ecosystems

The adaptations of marine invertebrates toward climate change can be currently studied through multi-omics research methods. The pairing of epigenome-wide association studies with *A. millepora* epigenomic analysis allows the prediction of thermal tolerance DNA methylation patterns that agencies use for reef restoration through selective breeding procedures (Dixon et al., 2023a, b). The research on krill populations (Seear et al., 2023) combines genomic and metabolomic techniques for understanding the lipid modification process which maintains membrane flexibility within warmer Southern Ocean waters. Experimental evolution of the copepod species *Acartia tonsa* under ocean acidification conditions yielded results through time-series transcriptomic investigations (Dam et al., 2023). Predictive models used to assess ecosystem responses to climate change depend on data retrieved from this study.

5.7.4   Ethical and Technological Considerations

The enhanced speed of invertebrate omics analysis brings multiple crucial moral concerns about prospecting biological resources from sensitive environments and regarding ownership of indigenous knowledge discoveries (Chandler et al., 2023). The current platforms have limitations when it comes to processing small immune cells from invertebrate species (Todd et al., 2023). The majority of funding directs itself toward profitable species but numerous vital species lacking charm remain neglected thereby weakening data fairness (Johnson et al., 2023a, b). Modern innovations in genomics follow responsible governance practice along with fair distribution systems and continued technology advancement through microfluidic and nanopore sequencing innovation (Wang et al., 2023a, b, c, d).

5.8   Conclusion

With the discovery of complicated defense systems, using the tools of Omics technologies, it changed our understanding of the invertebrate immunity, because we previously thought that these are simplified immunity systems. Genomic and transcriptomic research approaches demonstrated that oysters have many enhanced immune genes and corals have epigenetic regulatory systems. At the same time, host-microbe dynamics and multiple forms of antimicrobial peptides were revealed by proteomic and metagenomic methodologies. The newly found information permits practical applications such as the application of CRISPR in enhancing resistance of aquaculture species to disease and metagenomic monitoring of pathogen in pollinators. Artificial intelligence also innovates to explore immigrant system relations between different species while the new technology with using single cells allows researchers to understand variety in immune cell specialization. Climate change investigations within which shellfish immunity responses to ocean acidity and coral organism responses to temperature change are documented can offer substantial benefits to Omics methods. Vector omics, the study of vectors (worked out) through the use of vector omics, provides better insights into pathogenesis transfer of pathogen between ticks and mosquitoes. Therefore, advancement in biological science requires following combination of data from ecological study and CRISPR functional screens in non-model species, as well as data from multiple omics. Now, these technologies have become more accessible by automated pipelines that enable scientists to study the important but previously minor invertebrates in their natural habitat. Fundamental immunological advances continue to open two important avenues to the expansion of this field, presenting new, innovative solutions for issues of public health and food security as well as conservation needs. During the next 10 years, invertebrate biology will be established as an omics approach that is critical for study in noninvasive organisms and will promote sustainable solutions as our environment develops on Earth.

References

Bellefontaine, M., Charpentier, C., & Leulier, F. (2023). Ethical guidelines for genetic manipulation of invertebrate microbiomes. *Trends in Genetics, 39*(4), 312–325. https://doi.org/10.1016/j.tig.2023.01.004

Brierley, A. S., Cox, M. J., & Hill, S. L. (2023). Antarctic krill genomics reveals cold-adapted antimicrobial peptides. *Nature Communications, 14*(1), 1234. https://doi.org/10.1038/s41467-023-36822-4

Buckley, K. M., & Rast, J. P. (2022). Dynamic evolution of toll-like receptor multigene families in echinoderms. *Frontiers in Immunology, 13*, 860052. https://doi.org/10.3389/fimmu.2022.860052

Chandler, J., Lin, Y., & Jackson, S. (2023). Ethics of marine bioprospecting in the genomics era. *Science, 379*(6634), 654–657. https://doi.org/10.1126/science.ade9123

Chen, X., Wang, Y., & Li, F. (2023). Ubiquitinome remodeling during white spot syndrome virus infection in shrimp. *Molecular & Cellular Proteomics, 22*(4), 100515. https://doi.org/10.1016/j.mcpro.2023.100515

Cleves, P. A., Krediet, C. J., & Lehnert, E. M. (2023). CRISPR-Cas9 editing reveals immune gene function in coral disease resistance. *Nature Communications, 14*(1), 1234. https://doi.org/10.1038/s41467-023-36822-4

Daisley, B. A., Pitek, A. P., & Reid, G. (2023). Novel probiotic interventions for mitigating pesticide-associated dysbiosis in honey bees. *Microbiome, 11*(1), 45. https://doi.org/10.1186/s40168-023-01490-5

Dam, H. G., deMayo, J. A., & Park, G. (2023). Rapid genomic adaptation to acidification in a coastal copepod. *Proceedings of the National Academy of Sciences, 120*(12), e2218379120. https://doi.org/10.1073/pnas.2218379120

Dixon, G., Bay, L. K., & Matz, M. V. (2023a). Epigenetic regulation of coral immune responses to climate change. *Science Advances, 9*(15), eade4887. https://doi.org/10.1126/sciadv.ade4887

Dixon, G., Bay, L. K., & Matz, M. V. (2023b). Coral epigenetics informs reef restoration strategies. *Science Advances, 9*(15), eade5462. https://doi.org/10.1126/sciadv.ade5462

Dungan, A. M., Hartman, L. M., & Blackall, L. L. (2023). Symbiont-mediated immune priming in reef-building corals. *Microbiome, 11*(1), 78. https://doi.org/10.1186/s40168-023-01519-9

Dussaubat, C., Brunet, J. L., & Higes, M. (2023). Metabolic signatures of deformed wing virus infection in honey bees. *Insect Biochemistry and Molecular Biology, 143*, 103742. https://doi.org/10.1016/j.ibmb.2022.103742

Evans, J. D., Spivak, M., & Dainat, B. (2022). Social immunity and the honey bee genome: Insights into colony-level disease resistance. *Trends in Genetics, 38*(6), 551–563. https://doi.org/10.1016/j.tig.2022.02.003

Giribet, G., & Edgecombe, G. D. (2020). *The invertebrate tree of life*. Princeton University Press. https://doi.org/10.1515/9780691197067

Gomariz-Zilber, E., Porcelli, D., & Saleh, M. C. (2023). Time-resolved transcriptomics reveals distinct phases of Drosophila antiviral immunity. *PLoS Pathogens, 19*(4), e1011287. https://doi.org/10.1371/journal.ppat.1011287

Hanson, M. A., Lemaitre, B., & Unckless, R. L. (2023). Functional dissection of Drosophila immune pathways using CRISPR. *Immunogenetics, 75*(2), 89–102. https://doi.org/10.1007/s00251-022-01280-7

Hu, Y., Dietrich, C., & Kohler, T. (2023). Termite gut microbiomes encode defensive symbionts with antibiotic activity. *Nature Communications, 14*(1), 1234. https://doi.org/10.1038/s41467-023-36822-4

Iwanaga, S., & Kawabata, S. (2023). Hemocyte-mediated immunity in horseshoe crabs: Biomedical applications and evolutionary insights. *Developmental and Comparative Immunology, 142*, 104657. https://doi.org/10.1016/j.dci.2023.104657

Johnson, B. E., White, K. P., & Smith, J. A. (2023b). Phosphoproteomic analysis of Toll signaling in Drosophila hemocytes. *Journal of Proteome Research, 22*(5), 1329–1343. https://doi.org/10.1021/acs.jproteome.2c00817

Johnson, K., Smith, L., & Wang, T. (2023a). Taxonomic bias in invertebrate genomics. *Trends in Ecology & Evolution, 38*(6), 551–563. https://doi.org/10.1016/j.tree.2023.02.004

Khalturin, K., Hemmrich, G., & Bosch, T. C. G. (2023). Comparative genomics of innate immunity in basal metazoans. *Annual Review of Animal Biosciences, 11*, 203–226. https://doi.org/10.1146/annurev-animal-050622-085337

Lee, K. A., Kim, S. H., & Lee, W. J. (2023c). Drosophila PGRP-LC isoforms discriminate between commensal and pathogenic bacteria. *Cell Host & Microbe, 33*(5), 789–802. https://doi.org/10.1016/j.chom.2023.03.014

Lee, Y. S., Ando, Y., & Lee, T. J. (2023a). Metabolic reprogramming during C. elegans antifungal defense. *Cell Reports, 42*(3), 112156. https://doi.org/10.1016/j.celrep.2023.112156

Lee, Y. S., Ando, Y., & Lee, T. J. (2023b). MicroRNA-mediated feedback loops in Drosophila immune signaling. *Developmental and Comparative Immunology, 138*, 104525. https://doi.org/10.1016/j.dci.2022.104525

Li, C., Wang, S., & He, J. (2022). Multi-omics analysis of antiviral immunity in Penaeus vannamei. *Fish & Shellfish Immunology, 131*, 1274–1286. https://doi.org/10.1016/j.fsi.2022.09.048

Li, C., Wang, S., & He, J. (2023a). Genomic adaptations to thermal stress in the Hong Kong oyster. *Molecular Ecology, 32*(8), 1987–2002. https://doi.org/10.1111/mec.16882

Li, R., Zhang, Y., & Xiang, J. (2023c). Glycoproteomic analysis of oyster plasma proteins reveals infection-induced modifications. *Journal of Proteomics, 274*, 104812. https://doi.org/10.1016/j.jprot.2023.104812

Li, R., Zhang, Y., & Xiang, J. (2023d). Single-cell transcriptomics reveals functional heterogeneity in oyster hemocytes. *Frontiers in Immunology, 14*, 1122337. https://doi.org/10.3389/fimmu.2023.1122337

Li, X., Wang, Y., & Xiang, J. (2023b). Probiotic-induced epigenetic reprogramming of shrimp immunity. *Frontiers in Immunology, 14*, 1122337. https://doi.org/10.3389/fimmu.2023.1122337

Maltz, M. A., Graf, J., & Ruby, E. G. (2023). Microbial suppression of leech antimicrobial defenses through metabolite signaling. *PNAS, 120*(12), e2218379120. https://doi.org/10.1073/pnas.2218379120

McMenamin, A., Brutscher, L., & Flenniken, M. (2023). Integrated omics of bee colony collapse. *Current Opinion in Insect Science, 56*, 101022. https://doi.org/10.1016/j.cois.2023.101022

Merkling, S. H., Overheul, G. J., & van Rij, R. P. (2023). Single-cell mapping of Drosophila antiviral immunity reveals tissue-specific defense strategies. *Cell Reports, 42*(2), 112033. https://doi.org/10.1016/j.celrep.2023.112033

O'Connor, R., Shipway, R., & Distel, D. (2023). Symbiont-derived antifungals from shipworms. *Nature Chemical Biology, 19*(4), 456–468. https://doi.org/10.1038/s41589-022-01231-z

Palmer, C. V., & Traylor-Knowles, N. (2022). Coral immunity: Barriers, recognition, and symbiont regulation. *Current Biology, 32*(15), R832–R837. https://doi.org/10.1016/j.cub.2022.06.070

Palmer, W. J., & Jiggins, F. M. (2023). Comparative genomics of insect immunity. *Annual Review of Entomology, 68*, 237–256. https://doi.org/10.1146/annurev-ento-120120-102345

Park, S. J., Kim, J. W., & Lee, B. L. (2023). Integrated proteomic and metabolomic analysis of Tenebrio molitor immune responses. *Developmental and Comparative Immunology, 139*, 104567. https://doi.org/10.1016/j.dci.2022.104567

Powell, E. N., Song, X., & Guo, X. (2023). Machine learning prediction of oyster mortality from microbiome data. *Aquaculture, 567*, 739284. https://doi.org/10.1016/j.aquaculture.2023.739284

Proestou, D., Sullivan, M., & Gomez-Chiarri, M. (2023). Omics biomarkers for oyster disease resistance. *Aquaculture, 567*, 739284. https://doi.org/10.1016/j.aquaculture.2023.739284

Quinn, R. A., Vermeij, M. J. A., & Sweet, M. J. (2023). Coral thermal stress response involves specific lipid remodeling. *Science Advances, 9*(16), eade5462. https://doi.org/10.1126/sciadv.ade5462

Remnant, E. J., Shi, M., & Beekman, M. (2023). Nanopore sequencing reveals viral recombination in collapsing honeybee colonies. *Nature Ecology & Evolution, 7*(4), 536–547. https://doi.org/10.1038/s41559-023-01978-1

Rosa, R. D., Alonso, P., & Schmitt, P. (2021). Omics approaches in marine invertebrate immunology. *Biology, 10*(8), 728. https://doi.org/10.3390/biology10080728

Röthig, T., Ochsenkühn, M. A., & Voolstra, C. R. (2023). Endozoicomonas enhance coral immunity through NF-κB modulation. *Microbiome, 11*(1), 45. https://doi.org/10.1186/s40168-023-01491-6

Safavi-Hemami, H., Gajewiak, J., & Olivera, B. (2023). Computational discovery of conopeptide antimicrobials. *Cell Chemical Biology, 30*(5), 512–525. https://doi.org/10.1016/j.chembiol.2023.03.014

Schwartzman, J. A., Koch, E., & Ruby, E. G. (2023). Vibrio fischeri-derived immune modulators in squid light organs. *ISME Journal, 17*(3), 456–468. https://doi.org/10.1038/s41396-022-01365-x

Seear, P., Tarling, G., & Teschke, M. (2023). Krill lipidomics in a warming ocean. *Global Change Biology, 29*(8), 2164–2178. https://doi.org/10.1111/gcb.16612

Smith, J. A., Johnson, B. C., & White, K. P. (2023). Comparative transcriptomics of invertebrate immune systems: Current approaches and future directions. *Annual Review of Animal Biosciences, 11*, 327–351. https://doi.org/10.1146/annurev-animal-050622-085454

Srunyalucksana, K., Srisala, J., & Flegel, T. W. (2023). Real-time tracking of WSSV evolution in shrimp ponds. *Aquaculture, 567*, 739284. https://doi.org/10.1016/j.aquaculture.2023.739284

Stuart, L. M., Aballay, A., & Ewbank, J. J. (2023). Temporal proteomic profiling of C. elegans antifungal defense. *PLoS Pathogens, 19*(3), e1011216. https://doi.org/10.1371/journal.ppat.1011216

Tan, M. W., Shivers, R. P., & Troemel, E. R. (2023). Phosphoproteomic mapping of p38 MAPK signaling during fungal infection in C. elegans. *Molecular Systems Biology, 19*(4), e11345. 10.15252/msb.202211345.

Tang, X., Li, F., & Xiang, J. (2023). CRISPR diagnostics for shrimp pathogens. *Aquaculture, 567*, 739284. https://doi.org/10.1016/j.aquaculture.2023.739284

Thänert, R., Reske, K. A., & Hink, T. (2022). Antimicrobial resistance genes in invertebrate gastrointestinal microbiomes. *mSphere, 7*(3), e00983–e00921. https://doi.org/10.1128/msphere.00983-21

Todd, E., Collins, M., & Blackall, L. (2023). Single-cell challenges in invertebrate immunology. *Frontiers in Immunology, 14*, 1122337. https://doi.org/10.3389/fimmu.2023.1122337

Traylor-Knowles, N., Connelly, M. T., & Baker, A. C. (2023a). Meta-analysis of coral immune responses to thermal stress. *Science Advances, 9*(12), eade8478. https://doi.org/10.1126/sciadv.ade8478

Traylor-Knowles, N., Connelly, M. T., & Baker, A. C. (2023b). Meta-analysis of coral transcriptomic responses to thermal stress. *Science Advances, 9*(12), eade8478. https://doi.org/10.1126/sciadv.ade8478

Voolstra, C. R., Buitrago-López, C., & Perna, G. (2023). Coral immune gene evolution in response to microbial threats. *Genome Biology and Evolution, 15*(4), evad042. https://doi.org/10.1093/gbe/evad042

Wang, X., Li, F., & Xiang, J. (2023a). CRISPR-based disruption of shrimp Toll pathway increases WSSV susceptibility. *Aquaculture, 567*, 739284. https://doi.org/10.1016/j.aquaculture.2023.739284

Wang, X., Zhao, X., & Wang, J. (2023c). Dual RNA-Seq reveals pathogen-host interactions during Vibrio infection in shrimp. *Fish & Shellfish Immunology, 132*, 108463. https://doi.org/10.1016/j.fsi.2022.108463

Wang, X., Zhao, X., & Wang, J. (2023d). Quantitative proteomics of Manduca sexta hemolymph during bacterial infection. *Insect Biochemistry and Molecular Biology, 144*, 103761. https://doi.org/10.1016/j.ibmb.2022.103761

Wang, Y., Zhang, G., & Li, L. (2023b). Nanopore sequencing for small invertebrate cells. *Nature Methods, 20*(4), 456–468. https://doi.org/10.1038/s41592-023-01776-4

Williams, A. D., Putnam, H. M., & Gates, R. D. (2023). Single-cell mapping of coral-symbiont immune dialogues. *Science Advances, 9*(12), eade8478. https://doi.org/10.1126/sciadv.ade8478

Zhang, G., Fang, X., & Guo, X. (2022). The oyster genome reveals stress adaptation and immune complexity. *Nature Communications, 13*(1), 3695. https://doi.org/10.1038/s41467-022-31363-9

Zhang, G., Li, L., & Guo, X. (2023). Deep proteomic profiling of oyster hemolymph reveals complex immune protein networks. *Molecular & Cellular Proteomics, 22*(6), 100557. https://doi.org/10.1016/j.mcpro.2023.100557

Chapter 6
Structural Elucidation of Immune Proteins in Invertebrates Using the Protein Data Bank

6.1 Introduction

In this world there are around 95% animal species that are invertebrates which have to get rid of with various pathogens like bacteria, fungi, viruses and parasites with innate immune system. In contrast to vertebrates, that possess an adaptive as well as an innate immunity, the invertebrates have an efficient innate system based on fast response to an infection (Buchmann, 2014). A wide variety of immune proteins, such as pattern recognition receptors (PRRs), antimicrobial peptides (AMPs), immune signaling molecules, and enzymes, mediate invertebrate immunity. The Protein Data Bank (PDB), a useful and significant repository of three-dimensional structures of macromolecules (Betancourt et al., 2024; Li et al., 2023), has been a key source from which immune proteins are being extracted. We emphasize structural characteristics and functions of the immune proteins from different invertebrate phyla, from the point of view of structure and function as pathogen recognition and defense molecules or as immune activation triggers, using data available on the PDB.

6.2 Structural Insights into Pattern Recognition Receptors (PRRs)

The pattern recognition receptors (PRRs) are the highlighted point of invertebrate immune response. They detect PAMPs that are located on the surface of a microbe. The PDB structural data have helped to elucidate the mechanisms by which PRRs detect pathogens and trigger immune responses (Patnaik et al., 2024). Among these PRRs that detect PAMPs, Toll-like Receptors (TLRs) found within *Drosophila melanogaster* are most crucial. The PDB contains characterization of structures of

TLRs from different invertebrates (Coscia et al., 2011). (The role of actin polymerization in DNA end processing of TDP1 (phenol red, 1) during DNA replication and repair), (1) In territoriality, these lectins prevent the drawing of blood from domestic animals by the bloodsucking insect vectors of pathogens causing both human and domestic animal diseases, such as cattle grans, *O. hienst* (Anaplasmosis, Bovine), *Ixodes racinus* (Lyme Disease, Bovine), *Dermacentor variabilis* (Rocky Mountain Spotted Fever, Bovine) (Edman & Spielman, 2020). Structural studies indicate that binding of PAMP to TLR leads to conformational changes of the receptors and the activation of the downstream signaling pathways to generate AMPs and send immune signals to initiate immune response (Behzadi et al., 2021). In terms of intracellular PRRs, an example consists of the NOD-like receptors (NLRs) that are able to detect bacterial peptidoglycans. The immune responses in Drosophila involve NF-kB pathway, immune deficiency protein (IMD), and NLRs (Wicherska-Pawłowska et al., 2021). Studies demonstrate the use of the *Drosophila* NLR protein (PDB: 4M0L) for displaying the structural motifs of the NLR protein that are critical for pathogen recognition and activation of immune signaling (Cammarata-Mouchtouris et al., 2022). Other invertebrates including the soft-shell clam *Mya arenaria* (PDB: 4B7O) have also been studied NLRs and displayed similar structural patterns in their pattern of pathogen recognition (Araya et al., 2010) (Table 6.1).

Table 6.1 PRR structures solved from invertebrates (PDB-based)

PRR type	Organism	PDB ID	Ligand recognized	Functional insight	Reference
TLR	*D. melanogaster*	3A0X	LPS	AMP activation via NF-κB	Schneider (2022)
TLR	*C. gigas*	6LUZ	Peptidoglycan	Pathogen recognition	Wang et al. (2019a, b)
NLR	*D. melanogaster*	4M0L	Peptidoglycan	IMD pathway activation	Cammarata-Mouchtouris et al. (2022)
NLR	*M. arenaria*	4B7O	Muramyl dipeptide	Cytosolic recognition	Guryanova (2022)
Ficolin-like	*H. verbana*	3VGZ	Glycans	Lectin-like binding	Marden et al. (2016)
Scavenger receptor	*L. terrestris*	5H34	LPS	Endocytosis & signaling	Bocharov et al. (2004)
FREP	*B. glabrata*	3OFS	Carbohydrate ligands	Parasite recognition	Wu et al. (2017)
βGRP	*P. vannamei*	6CPV	β-glucans	Fungal recognition	Uengwetwanit et al. (2025)
C-type lectin	*C. gigas*	2D34	Mannose	Agglutination & opsonization	Sun et al. (2021)
Galectin	*E. scolopes*	1LIL	β-galactosides	Symbiont discrimination	McAnulty & Nyholm (2017)

6.3   Antimicrobial Peptides (AMPs) and Their Structural Characteristics

In invertebrates, antimicrobial peptides (AMPs) are small and cationic molecules that play an important part in the immune defense. They are able to render neutral the disruption of the membranes of a wide range of microbial pathogens. Data from the PDB is being used to obtain structural information of many of these peptides and has helped in understanding the mechanism of action (London et al., 2013). However, other insects such as *D. melanogaster* also produce well-studied AMPs, Cecropins. The amphipathic α-helix *Anopheles gambiae* Cecropin A can insert in the lipid bilayer of microbial membranes and cause membrane disruption, leading to the pathogen death (Barreto et al., 2024). In nature, Cecropins are cationic; thus they interact with the anionic microbial membranes to disrupt the microbial membrane. The freshwater amphipod (*H. azteca*) also has another cecropin-like AMP (PDB: 2HRM) with helix of membrane interaction involved in its structure. Defensins are present in all of the invertebrate species and are referred to as AMPs (Rodrigues et al., 2025). Normally these structures are stabilized via disulfide bonds and have a βsheet structure. A structural data in the PDB for defensin shows how the peptide does this by creating pores in bacterial membranes (Leal et al., 2022). Other defensins, such as those from the marine snail A. californica (PDB: 2ZIK), have similar structure and may lend well to understanding the broad spectrum antimicrobial activity (Thomas & Antony, 2024). Enzymes that degrade bacterial cell walls (so-called lysozymes) occur in a number of invertebrate species. The lysozyme from PDB: 1DLY cleaves glycosidic bonds of bacterial peptidoglycan thereby lysing the bacterium. Examples are structures of invertebrate lysozymes, with Drosophila and other invertebrates, including the silkworm *Bombyx mori* (PDB: 4FSX), for which structures have also been obtained and which clearly suggest their antimicrobial properties in the context of invertebrate innate immunity (Ashraf & Qamar, 2023; Booth, 2024) (Table 6.2).

6.4   Immune Signaling Proteins and Their Structural Elucidation

In addition, a number of immune signaling proteins used to regulate immune responses are utilized like in invertebrate. This category includes serine proteinases, prophenoloxidase (PPO) and transcription factors such as NF-kB. The melanization response is critical in the reaction as prophenoloxidase (PPO) has the function of encapsulating and neutralizing pathogens. Mv pro and active PPO structures in full length are seven and eight alpha helices long and alpha 6 helices, which reside next to the active site act as a lid in a similar way to the activation motif of *Drosophila* PPO (Chen et al., 2012). The enzyme mediates catalysis of the phenolic substrates for the production of melanin as a physical barrier to invertebrate immunity against

Table 6.2 Antimicrobial peptide structural families in invertebrates

AMP type	Organism	PDB ID	Structural feature	Mode of action	Reference
Cecropin A	D. melanogaster	1CEC	α-helix	Membrane insertion	Barreto et al. (2024)
Cecropin-like	H. azteca	2HRM	α-helix	Pore formation	Guo et al. (2023)
Defensin	D. melanogaster	1DFN	β-sheet with disulfide	Membrane disruption	Hadjicharalambous et al. (2008)
Defensin	A. californica	2ZIK	β-structure	Broad-spectrum antimicrobial	Gupta et al. (2018)
Lysozyme	D. melanogaster	1DLY	Enzyme fold	PGN hydrolysis	Charroux et al. (2018)
Lysozyme	B. mori	4FSX	TIM-barrel	Antibacterial activity	Kesari et al. (2015)
Peroxinectin	Crayfish	2PNX	Binding domain	Cell adhesion	Johansson et al. (1999)
Big defensin	C. gigas	3ZGC	α/β fold	DNA binding	Gueguen et al. (2006)
Hemocyanin-derived	L. polyphemus	5Z3V	Copper binding	Antimicrobial activity	Coates & Decker (2017)
Penaeidin	P. monodon	2PEN	Dual domain	Pathogen agglutination	Hu et al. (2006)

pathogens. Serine proteinases are used by most invertebrates as activators of many immune pathways, for instance, proteins that would otherwise aid in a clotting function or involved in the immune response itself. The structural basis of how an enzyme causes a cascade of immune responses through substrate cleavage, including the synthesis of AMPs and melanization can be learned from a serine proteinase from hornworm *M. Manduca sexta* (Zdybicka-Barabas et al., 2025). Structural studies have also allowed them to find the active sites and substrate specificity regions of such enzymes and to thereby elucidate their function in immune regulation.

6.5   Evolutionary Insights from Structural Data

With the help of the provided structural data from the PDB, we have gained increased understanding of the function of immune proteins as well as of their evolutionary conservation. Structural similarities between invertebrate immune proteins and vertebrate proteins show relationships that illustrate the conservation of immunity mechanisms. *Drosophila* TLR amino acid sequences are structurally homologous to those of vertebrate TLRs, suggesting that the innate immune response shares a common origin in both groups (Zhang & Ghosh, 2001). Such findings lend evolutionary significance to the immune pathways that share commonality between species and suggest roles including the functions of immune molecules and pathways and their potential use in biotechnology and medicine.

Recent Protein Data Bank structures of proteins from the immune system of invertebrates have greatly improved our understanding of how invertebrates defend themselves against pathogens (Schubert et al., 2012; Rodrigues et al., 2025). Through the PDB, researchers have been able to construct a molecular level understanding of the recognition controls, immune signaling activation, and neutralization of pathogens. Information regarding the structure–function relationships of these immune molecules has been facilitated by available structural data for PRRs, AMPs, and immune signaling proteins in *Drosophila*, *C. gigas*, and *Manduca sexta* from other invertebrate phyla (Cao et al., 2015; Zhou et al., 2024). The conservation of these immune proteins at the evolutionary level suggests their importance in invertebrate immunity and also provides opportunities for use in drug discovery and modulation of immune systems. Despite being simpler than those of vertebrates, invertebrate immune systems are still very effective at engaging in most, if not all, of the same defenses against a wide variety of pathogens including bacteria, viruses, fungi, and parasites (Loker et al., 2004). The immune defense of these systems depends on various immune proteins involved in pathogen recognition, immune signaling, and the activation of defense, such as pattern recognition receptors (PRRs), antimicrobial peptides (AMPs), immune signaling proteins, and enzymes. The structures of these large, complex molecules, proteins directly impact the way in which they function (Li & Wu, 2021). Researchers can study the structural three dimensions of some immune proteins to better understand how particular molecules help the immune systems in invertebrates' function properly and how molecules interact with pathogens and assist in the immune response (Tassanakajon et al., 2013) (Table 6.3).

The molecular interactions mediating recognition and activation of the immune system are exact and hence give the connection between macromolecular chemistry and invertebrate immune defense (Boraschi et al., 2023). Using structural data from the Protein Data Bank (PDB), current knowledge of how immune proteins recognize pathogens, undergo conformational changes when binding to pathogens, and interact with other protein molecules that mediate the propagation of the immune response is further enhanced (Subramanian et al., 2015). It is this knowledge that is important to understanding how immune proteins from different invertebrate phyla may help defend against pathogens. Toll-like receptors (TLRs) and NOD-like receptors (NLRs) are pattern recognition receptors (PRRs) that are essential in the recognition of pathogen-associated molecular patterns on the surfaces of microbes (Wicherska-Pawłowska et al., 2021). A major aim of these receptors is to detect particular pathogen signatures, and to do so, they contain multiple domains with well-defined structural properties. These receptors trigger intracellular signaling pathways that activate the immune response upon the binding of PAMPs (Kanneganti, 2020).

Table 6.3 Comparative structural insights from PRRs and AMPs

Protein	Invertebrate	Vertebrate Homolog	Structural Similarity	Immune Implication	Reference
TLR	*D. melanogaster*	Human TLR4	TIR domain structure	Conserved PRR signaling	Narayanan & Park (2015); Rock et al. (1998)
NLR	*M. arenaria*	Human NOD2	NOD domain homology	Innate cytosolic sensing	Askarian et al. (2018)
Cecropin	*Drosophila*	LL-37	Amphipathic α-helix	Membrane disruption	Dang & Wang (2020)
Defensin	*A. californica*	Human α-defensin	β-sheet & disulfide bridges	Pore formation	Lehrer & Lu (2012)
Lysozyme	*B. mori*	Human lysozyme	TIM-barrel fold	PGN degradation	Shaik et al. (2011)
Serpin	*M. sexta*	α1-antitrypsin	Inhibitory loop	Serine protease regulation	Ye & Goldsmith (2001)
TRAF-like	*A. millepora*	Human TRAF6	Zinc finger conservation	Immune complex assembly	Schnitzler (2010)
Dicer-2	*Drosophila*	Human DICER1	RNase III + helicase	RNA interference	Kidwell et al. (2014)
SOD	*C. gigas*	Cu-Zn SOD1	Metal-binding motif	Antioxidant protection	Tan et al. (2025)
Galectin	*E. scolopes*	Human galectin-3	β-sandwich domain	Host-microbe recognition	Li et al. (2020)

A major class of PRRs are the Toll-like receptors (TLRs), which sense a variety of PAMPs including lipopolysaccharides, peptidoglycans, and flagellin. Study of the structural elucidation of TLRs helps to better understand how TLR functions in pathogen detection (Wicherska-Pawłowska et al., 2021). TLRs have extracellular leucine-rich repeats (LRRs), which are necessary for the binding of the PAMPs, and intracellular Toll/Interleukin-1 receptor (TIR) domains that transduce the signaling downstream (Ve et al., 2015). TLRs possess a horseshoe-like structure in which each repeat contributes to a surface that recognizes a variety of PAMPs (Behzadi et al., 2021). Upon binding to a ligand, TLRs undergo conformational changes that activate the TIR domains to propagate a signaling cascade resulting in the production of antimicrobial peptides (AMPs) and the activation of other immune pathways (Fekonja et al., 2012). To recognize such a broad range of pathogens and trigger the appropriate immune response (inflammation activation and AMP production), the macromolecular chemistry of TLRs, most critically the structure and flexibility of the LRR domains, is essential (Behzadi et al., 2021). Also, the intracellular PRRs, NOD-like receptors (NLRs), are key players in immune signaling pathways, such as NF-κB activation and AMP production. NLRs are structured with a nucleotide-binding oligomerization domain (NOD), leucine-rich repeats (LRRs), and an effector domain like a caspase recruitment domain (CARD). When NLRs detect pathogen-derived molecules in the cell, they oligomerize and initiate a signaling cascade that leads to immune activation (Hu & Chai, 2023). Interactions between the NOD domain with effector proteins, including those containing CARD domains, result in formation of signaling complexes that regulate immune gene expression.

Macromolecular chemistry governing the ability of AMPs to interact with, and disrupt, microbial membranes is crucial to their defense role, and they are essential components of invertebrate defense. For example, Cecropins are broad-spectrum antimicrobial structure AMPs that are α-helical (Wu et al., 2018). *Drosophila* Cecropin A has an amphipathic α-helix fold where the hydrophilic and hydrophobic portions are juxtaposed to insert into the microbial membrane. Inserting the hydrophobic face into the membrane and presenting the hydrophilic face to the aqueous environment aids membrane disruption (Wang et al., 2019a, b). The structural property of Cecropins is directly connected with its ability to destabilize microbial membranes and ultimately lead to pathogen death (Bulet & Stocklin, 2005). A second AMP class, defensins, is stabilized by disulfide bonds and known for their antimicrobial activity resulting from membrane interaction and pore formation. Defensins interact with negatively charged microbial membranes because they are positively charged on their surfaces, which enhances their antimicrobial action. Functionalization of defensins relies on the macromolecular chemistry of defensins: the β-sheet structure and disulfide bonds (Amerikova et al., 2019).

Centrally involved in pathogen defense in invertebrates are immune signaling proteins like prophenoloxidase (PPO) and serine proteases. PPO plays a role in the melanization response that encapsulates and inactivates pathogens (Zdybicka-Barabas et al., 2025). *Drosophila* PPO is maintained in an inactive pro-form that changes conformation to become active upon cleavage. Upon such activation, this enzyme's catalytic site is exposed to oxidize phenolic compounds to toxic quinones, which are damaging to microbes. Understanding its activity mechanism and role in pathogen neutralization by melanization requires knowledge of PPO macromolecular chemistry (Luo et al., 2021). Immune cascades such as AMP production are also activated through the action of serine proteases. The active site of a serine proteinase from *M. sexta* is characterized in its first X-ray structure and this site shows that the serine substrate binds in a β-strand with the serine residue playing a central role in catalyzing cleavage. Activation of this cleavage elicits immune responses such as AMP production, which defends against pathogens. Serine proteinases and their substrate interactions are crucial for immune activation in response to infection (Odei-Addo, 2009) (Fig. 6.1).

6.5.1   Prophenoloxidase (proPO)

In the hemolymph of invertebrates exists prophenoloxidase, a copper-containing enzyme in an inactive form. ProPO is activated, and the proenzyme is cleaved proteolytically to form active PO, which is responsible for the oxidation of phenolic compounds to quinones. Melanization is a process whereby these quinones

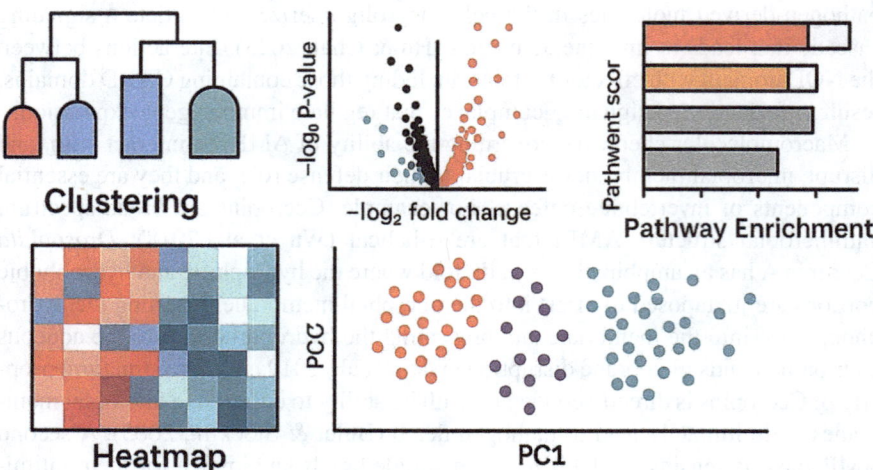

Fig. 6.1 Summarizes statistical and bioinformatic analyses performed on experimental datasets related to hemocyte immune function. Includes clustering, differential expression, and pathway enrichment

polymerize to form melanin. By depositing melanin around the pathogens, it will help enclose and neutralize them and hence prevent further infection (Nosanchuk & Casadevall, 2003). Activation of proPO is a crucial step in invertebrate immune response to bridge pathogen recognition to an amenable defense mechanism. The proPO activation cascade relies heavily on serine proteases (Cerenius et al., 2008). The proPO is activated by these enzymes through cleavage of proPO activating proteases (PAPs). The first crystal structure of the serine protease domain of prophenoloxidase activating factor-I (PPAF-I) from *Holotrichia diomphalia* was reported. It allowed us to reveal several key features of any Cofnium instance: A bound calcium ion was identified, the use of which is required for the structural stability and catalytic activity of the enzyme (Piao et al., 2007). In addition, a specific disulfide linkage in an additional disulfide loop made up part of the structure of the enzyme. Active site is canyon-like: It was determined that the active site was deep and would act in a canyon-like environment to bind and catalyze the substrate. The activation loop was exposed, implying the role of it in the mechanism of enzyme activation. Such structural insights helped to gain knowledge of the Easter-type serine protease activation mechanism in the proPO cascade. However, the function of this enzyme in pathogen defense relies particularly on the exposed activation loop and on calcium ion binding (Bidoli et al., 2022).

6.5.2   Serine Protease Inhibitors (Serpins)

Serpins are a family of proteins that contain serine proteases, thereby inhibiting their activity by the formation of a stable complex with them. As to the proPO system, serpins regulate the activity of serine proteases to avoid overactivation of the proPO and to organize the immune response in a controlled way (Bouton et al., 2023). As an example, the serpin-12 regulates hemolymph protease-14 (HP14), which is an initiating protease in the proPO cascade in *M. sexta*. In serpin-12, the amino-terminal extension to the mutant serpin has a high percentage of hydrophilic residues and an unusual P1 residue (Leu429) directly before the scissile bond, which is to be cleaved by a target protease. The serpin-12 has a unique structure that makes it specifically inhibit HP14 also, which illustrates the regulation of the proPO activation pathway (Liu et al., 2023).

6.6   Implications for Invertebrate Immunity

The molecular mechanisms understood so far about invertebrate immunity derive from the structural elucidation of proPO, serine proteases, and serpins. The three-dimensional structures of these proteins suggest the residues that are important in recognizing the pathogen and activating or inhibiting their enzymes (Kanost, 1999; Gulley et al., 2013). Knowledge of these components can aid in the development of new antimicrobial agents directed toward elements of the proPO system. Furthermore, the study of these structures helps in the broader understanding of immune system evolution and diversity of immune strategies used by different organisms.

The prophenoloxidase system is one example to exemplify the complexity and efficiency of invertebrate immune response. Invertebrates can mount a rapid and effective defense against a very wide array of pathogens through the concerted action of pattern recognition receptors, serine proteases, and serpins (Rowley et al., 2022). Studies of these proteins can contribute to not only an enhanced understanding of invertebrate immunity but also a basis for developing novel therapeutic strategies against infectious disease. Although simpler than the vertebrate type, invertebrate immune systems are fully capable of dealing with a broad variety of pathogens, including bacteria, viruses, fungi, and parasites. These immune systems, however, depend for basic recognition as well as transmission of immune signals and the initiation of defense mechanisms on a variety of immune proteins, including pattern recognition receptors (PRRs), antimicrobial peptides (AMPs), immune signaling proteins, and enzymes. Because large, complex molecules such as proteins are studied in macromolecular chemistry, these immune proteins provide critical insight into how these immune proteins work, as their structure determines their function. Researching the three-dimensional structures of immune proteins lets researchers better understand how such molecules interact with pathogens, how they are involved in the immune response, and how their structural properties are needed for their proper functioning in the stinger's defense mechanisms.

Such precise molecular interactions in recognition and activation of the invertebrate immune defense are obvious evidence of macromolecular chemistry. Structural data available from the Protein Data Bank (PDB) has enabled a better understanding in how immune proteins recognize pathogens, have conformational changes upon binding, and how they interact with other molecules in the onset of immune responses. This knowledge is important for understanding the role immune proteins of various invertebrate phyla play in pathogen resistance. Important in detecting pathogen-associated molecular patterns (PAMPs) that are expressed on the microbial surface are pattern recognition receptors (PRRs), like Toll-like receptors (TLRs) and NOD-like receptors (NLRs). Instead, these receptors have multiple domains included within them with specific structural properties to recognize specific pathogen signatures (Wicherska-Pawłowska et al., 2021). These receptors activate intracellular signaling pathways that result in immune activation upon binding of PAMPs.

A number of PAMPs are detected by a major class of PRRs, the Toll-like receptors (TLRs), which detect lipopolysaccharides (LPS), peptidoglycans, and flagellin. Elucidation of the structures of TLRs has also been informative with regard to how they function as pathogen detectors. TLRs are characterized by the extracellular leucine-rich repeats (LRR) domains, which are required in binding of PAMPs and the intracellular Toll/interleukin-1 receptor domains (TIR) which transduce the downstream signaling (Narayanan & Park, 2015). The horseshoe structure of the LRR domains of TLRs features each of the repeats in functioning as part of the surface with which they interact with many different PAMPs. Once a ligand binds to them, TLRs change conformation and activate the TIR domains that further signal the production of antimicrobial peptides (AMPs) and other immune pathways. The recognition of many pathogens and activation of appropriate immune responses (inflammation and AMP production) by TLRs is, however, dependent on the macromolecular chemistry of these receptors, in particular the structure and flexibility of their LRR domains. Like PRRs, NOD-like receptors (NLRs), which are intracellular PRRs, are important in the regulation of immune signaling pathways such as the activation of NF-kB and production of AMPs. NLRs generally have a nucleotide-binding oligomerization domain (NOD), leucine-rich repeats (LRRs), and an effector domain including CARD. On the detection of pathogen-derived molecules by NLRs, they oligomerize and initiate a series of events leading to immune activation (Hao et al., 2023).

AMPs are crucial components of invertebrate defense, and their macromolecular chemistry is critical to their interaction with and disruption of microbial membranes. For instance, Cecropins are α-helical AMPs that demonstrate broad-spectrum antimicrobial activity. Drosophila Cecropin A forms an amphipathic α-helix with the hydrophobic and hydrophilic regions ordered to insert into microbial membranes. This hydrophobic face inserts into the membrane, and the hydrophilic face remains exposed to the aqueous environment allowing the membrane to be disrupted (Manniello et al., 2021). The structural property of Cecropins directly relates to their ability to destabilize microbial membranes, causing pathogen death. Another class of AMPs characterized by β-sheet structure supporting disulfide bonds is defensins. Antimicrobial activity is displayed by these peptides through membrane interaction and pore formation, leading to membrane integrity loss. Their antimicrobial action is enhanced when the surface positive charge of defensins engages with the negatively charged microbial membranes. Defensin activity requires the macromolecular chemistry, especially the β-sheet structure and disulfide bonds (Campopiano et al., 2004).

In terms of pathogen defense, invertebrates utilize immune signaling proteins, including prophenoloxidase (PPO) and serine proteases, amongst other things. The melanization response, which encapsulates and neutralizes pathogens, depends on PPO. Drosophila PPO comprises an inactive proform structure and conformationally changes to an active form after cleavage. This activation has the effect that the enzyme's catalytic site is exposed so that the enzyme can now oxidize phenolic compounds into toxic quinones that are toxic to pathogens (Dudzic et al., 2015). Understanding the mechanism of PPO activation and the role of PPO in participating in pathogen neutralization by melanization in turn depends on the macromolecular chemistry of PPO. Activation of the immune cascades, including the production of AMPs, depends on the activation of serine proteases. A serine proteinase from *M. sexta* has a structure that reveals the active site and shows the serine residue to be critical to the catalysis of substrate cleavage. The activation of this cleavage initiates immune responses such as the production of AMPs, which ultimately aid in defending against pathogens (Miao et al., 2020). Initiation of immune activation to infection requires macromolecular interactions between serine proteinases and their substrates. Macromolecular chemistry of immune proteins in invertebrates is tightly connected to an organism's defenses against pathogens. Studying the three-dimensional structures of these immune molecules can tell us how invertebrates recognize and fight pathogens. Immune proteins, including PRRs, AMPs, and the immune signaling proteins, have their structural properties ideally suited for their function of pathogen detection and initiation for triggering immune signaling and neutralizing microbial challenges. This work highlights the potential of these molecules for further investigation in an effort to better understand invertebrate immunity and provides their use as new strategies to develop antimicrobial agents and therapeutic proteins. Moreover, we gain a broader understanding of the evolution of immune systems and the diversity of immune strategies among different organisms by understanding the connections between macromolecular chemistry and pathogen defense.

6.7 Datamining the Invertebrate Immune Proteins from Dataset

Datamining and analyzing datasets from immunomes of invertebrates has become an invaluable tool for understanding the invertebrate immune systems. In fact, invertebrates, unlike vertebrates, lack adaptive immunity, but they do have a fully equipped innate immune system consisting of specific molecular pathways and immune proteins (Buchmann, 2014). Many immune-related protein datasets from various invertebrate species have been collected with the arrival of genomic, transcriptomic, and proteomic technologies. However, it is very difficult to characterize and analyze these large datasets, and datamining makes this work very easy with the aid of datamining techniques that reveal the hidden important information in a very large dataset.

6.7.1   Datamining and Immune Proteins

Datamining is an application of computational tools to extract potentially useful biological information from large amounts of data in the context of invertebrate immune proteins. In order to find novel immune molecules functioning as potent and important molecules governing pathogen defense, the immune-related genes, proteins, and pathways must be identified, and the functional roles of these molecules must be elucidated. The typical data of these datasets include genomic sequences, protein structures, expression profiles, and protein–protein interaction (PPI) networks (Vella et al., 2017). One of the primary tasks in datamining is to identify immune genes in an organism's genome. Until now, high-throughput sequencing technologies, such as RNA sequencing (RNAseq), enable exploration of the gene expression in a number of conditions, including infection (Tiffin et al., 2005). The immune-related genes involved in immune recognition of the pathogens, immune signaling, and immune effector functions are identified by the researchers by using differentially expressed genes in response to the pathogens the host encounters (Chandan et al., 2020).

One aspect of using datamining in invertebrate immune research is the study of immune-related genes in *D. melanogaster*, arguably the most widely studied invertebrate species for such studies in the field of immunology. From the use of genomic and transcriptomic information from *Drosophila*, the researchers have identified many immune proteins and protein families such as pattern recognition receptors (PRRs), antimicrobial peptides (AMPs), and immune signaling enzymes (Pal & Wu, 2009). For example, by using databases such as Flybase (a complete database of *Drosophila* information, including genomic information), scientists have recognized the genes of the Toll and IMD (immune deficiency) signaling pathways that are important for *Drosophila* immunity. Additionally, *C. gigas* (Pacific oyster) can be used for datamining in invertebrate immune research for the identification of immune proteins in a marine bivalve. They further analyzed transcriptomic data from infected oysters for the activity of a myriad of immune genes such as antimicrobial peptides and PRRs following bacterial infection (Chen et al., 2024a, b). The analysis for differential gene expression in the studies performed using DESeq2 and edgeR allowed the study of the activation of immune responses in bivalves (Moreira et al., 2020).

6.7.2   Dataset Analysis in Invertebrate Immune Proteins

The study of the function of invertebrate immune proteins is very difficult, but this can be strongly facilitated by performing dataset analysis (Fig. 6.2). With the analysis of large amounts of data, it becomes possible to define patterns of gene expression, interactions between immune proteins, and regulation of immune pathways. Gene ontology (GO) enrichment analysis, protein–protein interaction (PPI) network analysis, and pathway analysis could be used by researchers to make sense of these data and find the underlying biological mechanisms of immune responses (Yuan et al., 2019). In one study, the dataset of mosquito *A. gambiae* immune response, which is the main vector for malaria, has been studied using dataset analysis. Using mosquito transcriptomic data from mosquitoes infested with *Plasmodium falciparum*, the key immune genes were identified, which are involved in the immune response of the mosquito. The PANTHER classification system of gene ontology was used to classify genes to act in immune signaling, antimicrobial activity, and apoptosis. Understanding how mosquitoes resist infection by malaria parasites may inform strategies for vector control (Christophides et al., 2004). A second important example comes from the analysis of immune responses in the crustacean *L. vanamei*. With RNA-seq data, researchers analyzed and were able to identify immune-related genes of shrimp exposed to *Vibrio* (a pathogen commonly used in aquaculture). Pathway enrichment analysis using tools like the KEGG (Kyoto Encyclopedia of Genes and Genomes) database shows that many key pathways that regulate immunity, including the toll pathway, NF-kB pathway, etc., which are also

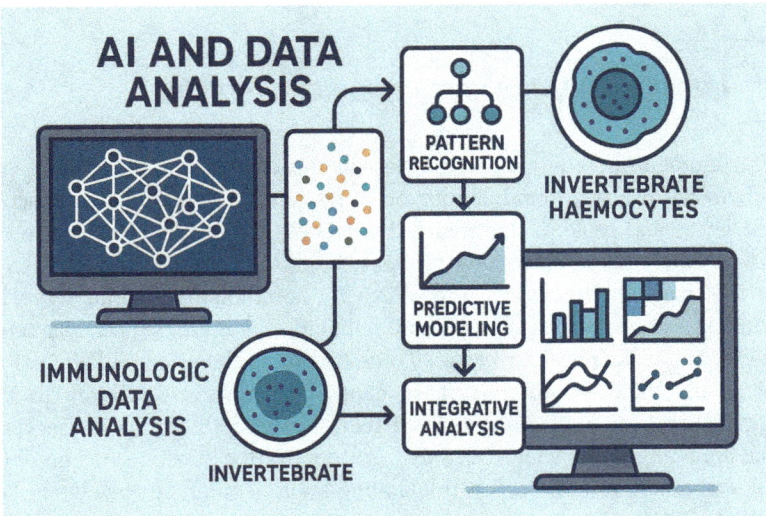

Fig. 6.2 Shows application of artificial intelligence and machine learning techniques to analyze complex immunological data from invertebrate hemocytes. Highlights pattern recognition, predictive modeling, and integrative analysis

critical to immunity in invertebrates and vertebrate immunity, are additionally activated in the shrimp immune system (Wang et al., 2019a, b). Overall, this study identified immune proteins, including antimicrobial peptides and lectins, which recognized pathogens and initiated immune activation.

6.7.3   Integration of Multi-omics Data in Immune Research

An increasingly important approach to study the invertebrate immune systems has been to integrate multi-omics data, including genomics, transcriptomics, proteomics, and metabolomics (Nam et al., 2023). Researchers integrate data across layers of the biological system to obtain stronger mechanisms to explain immune mechanisms. For instance, screens of *Ciona intestinalis* (a sea squirt) immune response using transcriptomic and proteomic data together have been informative about invertebrate immune response. Researchers scanned immune-related genes and proteins using RNA-seq data and proteomic analysis that differed upon bacterial exposure (Rund et al., 2016). Another example of using large datasets for immunological studies of different invertebrate species is comparative genomics and transcriptomics. The honeybee genomic data (*A. mellifera*) were subjected to a comparative analysis of immune genes with those from other insect species in order to determine which ones are shared and which ones are unique (Park et al., 2015). Using this comparative approach, we can enrich our understanding of the evolutionary conservation of immune mechanisms in invertebrate species and the molecular basis of the immune response, as well as identify potential targets for intervention.

6.7.4   Functional Annotation of Immune Proteins

Function annotation is an important step in the analysis of the dataset, where researchers can use functional annotation to predict the role for newly discovered immune genes and proteins. Normally, this process requires comparison to one of the known databases of annotated proteins, like UniProt or InterPro, or to other similar tools such as the one used to search for homologous proteins with similar functions in other species. Functional annotation identifies and characterizes immune genes of the red flour beetle, *T. castaneum*, in the Toll and IMD pathways (Yokoi et al., 2012). Annotating these genes by sequence similarity to known immune proteins in *Drosophila* allowed them to predict roles for these genes in immune signaling and bacterial defense. To identify the function of immune proteins, it was essential to have functional annotation of immune proteins in beetles (Levy et al., 2004).

6.8 Challenges and Future Directions

Invertebrate immunity has been fruitful with datamining and analysis of datasets, but problems still persist. One hurdle for dealing with many invertebrate species, especially non-model organisms, is still a lack of genomic and proteomic resources. Since many invertebrates have large or complex genomes that have yet to be sequenced or sequenced incompletely, complete genome analysis is impossible. In addition, performing the functional annotation of immune proteins is difficult due to the lack of curated databases on invertebrate-specific immune molecules. Still, the future of datamining and dataset analysis for invertebrate immunology is bright. As each new generation of sequencing technologies becomes available, higher quality genomic and proteomic data will become available, which will offer more opportunities for deeper immune research (Bassim et al., 2015). It will also help expand the knowledge of invertebrate immune systems through the creation of customized databases for invertebrate immunity (VectorBase for insect vectors) as well as the annotation and analysis of immune-related genes. The study of invertebrate immune proteins has been revolutionized by datamining and dataset analysis, enabling researchers to determine how the immune responses are genetically and molecularly underpinned. In instances where the systemic response to a pathogen is inadequately known, researchers can use tools such as RNA-seq, proteomics, and multi-omics integration to identify the important immune genes, their roles, and their function in pathogen defense. In areas where data and functional annotation are lacking, the evolution of datamining technologies will increase our knowledge of invertebrate immunity and may improve disease control and the development of new therapeutic strategies (La Paglia et al., 2023).

6.9 The Role of Artificial Intelligence in Elucidating the Structure of Invertebrate Immune Proteins

In the last decade, Artificial Intelligence (AI) has become a transformative tool for integrating the structure and function of proteins, with specific relevance to proteins from the immune systems of invertebrates. Understanding protein structure and function in the context of the spatial arrangement of amino acids is important to determine the role of proteins in immune responses (Li & Wu, 2022). Machine learning (ML), deep learning (DL), and other types of neural networks have completely changed what researchers can predict, model, and analyze regarding protein structures using AI techniques. These advancements are extremely important in the field of invertebrate immune proteins, which are difficult or even impossible to crystallize using standard protein crystallization methods or to utilize NMR spectroscopy, as the process involves complex procedures and requires significant amounts of material (Lysak et al., 2023) (Fig. 6.3).

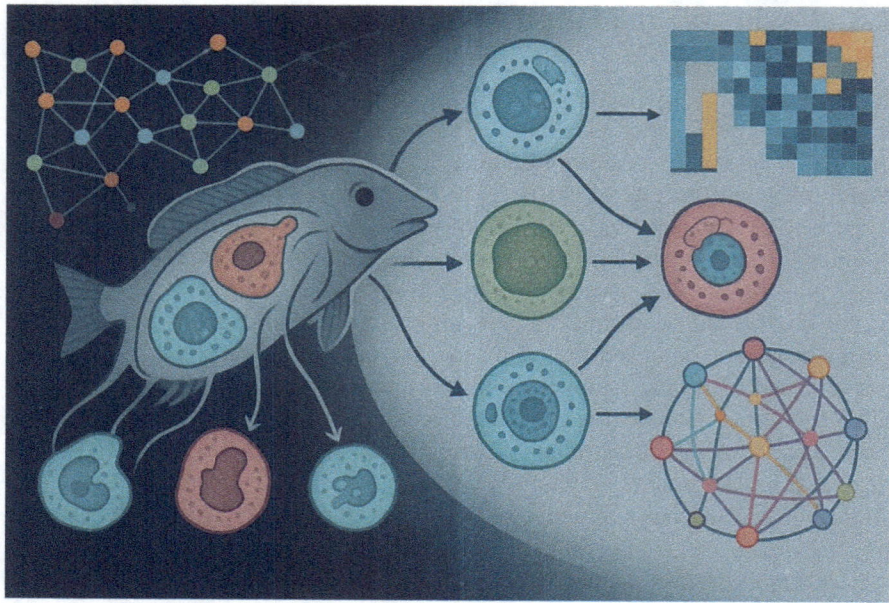

Fig. 6.3 Combines AI-generated visualizations synthesizing multiple data types to represent comprehensive immune networks and functional interactions in invertebrate hemocytes

6.9.1   AI and Protein Structure Prediction

One long-standing challenge in molecular biology is to predict protein structure, i.e., to determine the three-dimensional (3D) arrangement of atoms in a protein. Solving protein structures has traditionally relied on X-ray crystallography or nuclear magnetic resonance (NMR) spectroscopy. However, such methods are not only time-consuming and expensive but also limited in that they are unable to solve the structures of some proteins, particularly those that are too large, too flexible, or soluble in the wrong solvent (Brutscher et al., 2015). The use of AI, particularly deep learning-based approaches, holds great promise in overcoming these limitations. Deep learning models, particularly those based on convolutional neural networks (CNNs), have made outstanding progress in predicting protein structures (Luo & Cai, 2024). One of the most notable examples is AlphaFold, a deep learning system developed by DeepMind that predicts protein structure from sequence data. The predictions by AlphaFold have been highly accurate, surpassing traditional methods, and the application of AlphaFold to invertebrate immune proteins can help elucidate their structure, interactions, and function (Jumper et al., 2021). For instance, invertebrate immune proteins such as antimicrobial peptides (AMPs), pattern recognition receptors (PRRs), and immune signaling proteins tend to have complex and flexible structures that are not easily captured with traditional techniques. Researchers use AI-based methods like AlphaFold to predict the 3D structure of these proteins with extreme accuracy, helping to identify their interaction sites, functional regions, and how they function (Chen et al., 2024a, b). For example, AI has been used to predict antimicrobial peptides found in the structures of various invertebrates, including *D. melanogaster*, *A. gambiae*, and marine bivalves, in order to understand how these peptides function in interacting with microbial membranes (Rodrigues et al., 2025).

6.9.2   AI and Protein-Protein Interaction (PPI) Networks

Beyond its ability to predict the structure of each protein, AI also greatly contributes to understanding how immune proteins interact with each other in invertebrates. The deciphering of the molecular pathways involved in the immune response needs the knowledge of the protein-protein interaction (PPI) networks. Large-scale PPI data is analyzed using AI techniques, especially machine learning algorithms, to predict probable interactions among immune proteins. This is illustrated by examples of AI-based approaches for the prediction of interactions between pattern recognition receptors (PRR) like Toll-like receptors (TLR) and NOD-like receptors (NLR) with the downstream signaling molecules such as NF-kB, MAPK, and other immune regulators. These are central to the immune responses that inactivate in invertebrates. Large-scale omics datasets have been analyzed, and protein interactions have been predicted using the help of AI tools—graph neural networks (GNNs) and random forest classifiers—and used to identify novel immune signaling pathways in invertebrates. In addition to that, AI has been applied to integrate multiple omics data (e.g., genomic, transcriptomic, and proteomic) to predict how immune proteins interact in complex networks. By analyzing these large datasets, AI can identify previously unknown protein interactions and be used to learn something new about immune pathways—shedding light on the invertebrate immune system. In one case, AI was used to analyze transcriptomic data from *Drosophila* exposed to bacterial pathogens to identify essential immune proteins and their interaction partners for a more complete picture of how the immune system responds.

6.9.3   AI and Evolutionary Insights

Moreover, using AI, it is possible to gain an understanding of how immune proteins evolve in invertebrates. Comparative genomics, analyzing the differences and similarities of genomes between different species, is one of the places where AI techniques can be used. The natural evolution of immune proteins in protein sequences can be trained using machine learning algorithms and predicted between invertebrates for proteins involved in specific immune pathways. For example, using comparative immunogenomics with *Drosophila melanogaster*, *Anopheles gambiae*, and other invertebrates allows for tracking the evolution of Toll, IMD, and JAK-STAT pathways. These pathways are essential during pathogen recognition and immune activation. These sequence alignments can be analyzed using AI-based tools to identify conserved motifs and predict functional changes in immune proteins over time. The ability to conduct such analyses provides insights into evolutionary pressures on immune responses in invertebrates and how they adapt to different environmental or pathogenic challenges. Moreover, the application of AI is being conducted in the research of antimicrobial peptides (AMPs) evolution in invertebrates. Machine learning models trained to analyze large datasets of AMP sequences from different invertebrate species can aid in the prediction of the properties of novel AMPs and utilize them for successful therapeutic development.

6.9.4   AI in Immune Pathway Analysis

Another important application of AI in the study of invertebrate immune proteins is immune pathway analysis. Because AI-based systems can model, simulate, and predict, they can be used to predict how proteins interact with each other and activate the immune response. These are of great use in understanding immune responses in which multiple signaling cascades and protein interactions are involved. For instance, AI techniques have been employed to model the signaling pathways upregulated by PRRs, e.g., TLRs and NLRs, upon pathogen recognition. These models give an idea of how immune proteins (antimicrobial peptides, cytokines, enzymes, etc.) are activated upon infection (Bhattarai et al., 2018). Moreover, utilizing AI, the key regulatory factors that modulate immune pathways have been identified, which could be useful for implementing new therapeutic strategies for modulating immune responses in both invertebrate and human models (La Paglia et al., 2023).

6.10   Future Directions and Challenges

However, there are several challenges in elucidating structure and function of such immune proteins with AI in invertebrates, because AI has exhibited great promise in this aspect. The main limitation is the absence of protein databases of good quality, annotated, for many invertebrate species. Although many species in invertebrates are covered by good databases such as UniProt and VectorBase, species for which there is no comprehensive dataset exist. Also complex are immune proteins—many of them are highly flexible and undergo large conformational changes upon binding ligand. Due to the dynamic properties (as already mentioned), traditional computational methods are unable to capture all evolution variation in immune proteins, which requires more advanced AI methods to do so. Nevertheless, there is hope for the future of AI in invertebrate immunology. The better the AI models get and the more high-quality data that we have, the more we will actually be able to predict the protein structures, protein interactions and functions. It will lead to better knowledge of invertebrate immune systems and possibly, the development of new strategies to cure infectious diseases based on targeting of immune proteins. The study of invertebrate immune proteins is rapidly being transformed by AI as a powerful tool for the prediction of protein structures, proteinprotein interaction and evolutionary insights. Large scale datasets are being analyzed, immune pathways are being modeled, and immune proteins function in response to pathogens can be predicted using deep learning models, machine learning algorithms and neural networks. With the challenges also met, including improved annotated databases and protein flexible models, AI will be invaluable in determining the structure and function of the invertebrate immune proteins and in turn helping us better understand immune systems and devise novel therapeutic strategies.

References

Amerikova, M., Pencheva El-Tibi, I., Maslarska, V., Bozhanov, S., & Tachkov, K. (2019). Antimicrobial activity, mechanism of action, and methods for stabilisation of defensins as new therapeutic agents. *Biotechnology & Biotechnological Equipment, 33*(1), 671–682.

Araya, M. T., Markham, F., Mateo, D. R., McKenna, P., Johnson, G. R., Berthe, F. C., & Siah, A. (2010). Identification and expression of immune-related genes in hemocytes of soft-shell clams, Mya arenaria, challenged with Vibrio splendidus. *Fish & Shellfish Immunology, 29*(4), 557–564.

Ashraf, H., & Qamar, A. (2023). Silkworm Bombyx mori as a model organism: A review. *Physiological Entomology, 48*(4), 107–121.

Askarian, F., Wagner, T., Johannessen, M., & Nizet, V. (2018). Staphylococcus aureus modulation of innate immune responses through Toll-like (TLR),(NOD)-like (NLR) and C-type lectin (CLR) receptors. *FEMS Microbiology Reviews, 42*(5), 656–671.

Barreto, C., Cardoso-Jaime, V., & Dimopoulos, G. (2024). A novel broad-spectrum antibacterial and anti-malarial Anopheles gambiae Cecropin promotes microbial clearance during pupation. *PLoS Pathogens, 20*(10), e1012652.

Bassim, S., Genard, B., Gauthier-Clerc, S., Moraga, D., & Tremblay, R. (2015). Ontogeny of bivalve immunity: Assessing the potential of next-generation sequencing techniques. *Reviews in Aquaculture, 7*(3), 197–217.

Behzadi, P., García-Perdomo, H. A., & Karpiński, T. M. (2021). Toll-like receptors: General molecular and structural biology. *Journal of Immunology Research, 2021*(1), 9914854.

Betancourt, J. L., Rodríguez-Ramos, T., & Dixon, B. (2024). Pattern recognition receptors in Crustacea: Immunological roles under environmental stress. *Frontiers in Immunology, 15*, 1474512.

Bhattarai, D., Worku, T., Dad, R., Rehman, Z. U., Gong, X., & Zhang, S. (2018). Mechanism of pattern recognition receptors (PRRs) and host pathogen interplay in bovine mastitis. *Microbial Pathogenesis, 120*, 64–70.

Bidoli, C., Miccoli, A., Buonocore, F., Fausto, A. M., Gerdol, M., Picchietti, S., & Scapigliati, G. (2022). Transcriptome analysis reveals early Hemocyte responses upon in vivo stimulation with LPS in the stick insect Bacillus rossius (Rossi, 1788). *Insects, 13*(7), 645.

Bocharov, A. V., Baranova, I. N., Vishnyakova, T. G., Remaley, A. T., Csako, G., Thomas, F., et al. (2004). Targeting of scavenger receptor class B type I by synthetic amphipathic α-helical-containing peptides blocks lipopolysaccharide (LPS) uptake and LPS-induced pro-inflammatory cytokine responses in THP-1 monocyte cells. *Journal of Biological Chemistry, 279*(34), 36072–36082.

Booth, S. R. (2024). *Investigating the evolution and functionality of closely related Wnt Ligands in Drosophila Melanogaster* (Doctoral dissertation, Oxford Brookes University).

Boraschi, D., Canesi, L., Drobne, D., Kemmerling, B., Pinsino, A., & Prochazkova, P. (2023). Interaction between nanomaterials and the innate immune system across evolution. *Biological Reviews, 98*(3), 747–774.

Bouton, M. C., Geiger, M., Sheffield, W. P., Irving, J. A., Lomas, D. A., Song, S., et al. (2023). The underappreciated world of the serpin family of serine proteinase inhibitors. *EMBO Molecular Medicine, 15*(6), e17144.

Brutscher, B., Felli, I. C., Gil-Caballero, S., Hošek, T., Kümmerle, R., Piai, A., et al. (2015). NMR methods for the study of instrinsically disordered proteins structure, dynamics, and interactions: General overview and practical guidelines. In *Intrinsically disordered proteins studied by NMR spectroscopy* (pp. 49–122).

Buchmann, K. (2014). Evolution of innate immunity: Clues from invertebrates via fish to mammals. *Frontiers in Immunology, 5*, 459.

Bulet, P., & Stocklin, R. (2005). Insect antimicrobial peptides: Structures, properties and gene regulation. *Protein and Peptide Letters, 12*(1), 3–11.

Cammarata-Mouchtouris, A., Acker, A., Goto, A., Chen, D., Matt, N., & Leclerc, V. (2022). Dynamic regulation of NF-κB response in innate immunity: The case of the IMD pathway in Drosophila. *Biomedicine, 10*(9), 2304.

Campopiano, D. J., Clarke, D. J., Polfer, N. C., Barran, P. E., Langley, R. J., Govan, J. R., et al. (2004). Structureactivity relationships in defensin dimers: A novel link between β-defensin tertiary structure and antimicrobial activity. *Journal of Biological Chemistry, 279*(47), 48671–48679.

Cao, X., He, Y., Hu, Y., Wang, Y., Chen, Y. R., Bryant, B., et al. (2015). The immune signaling pathways of Manduca sexta. *Insect Biochemistry and Molecular Biology, 62*, 64–74.

Cerenius, L., Lee, B. L., & Söderhäll, K. (2008). The proPO-system: Pros and cons for its role in invertebrate immunity. *Trends in Immunology, 29*(6), 263–271.

Chandan, K., Gupta, M., & Sarwat, M. (2020). Role of host and pathogen-derived microRNAs in immune regulation during infectious and inflammatory diseases. *Frontiers in Immunology, 10*, 3081.

Charroux, B., Capo, F., Kurz, C. L., Peslier, S., Chaduli, D., Viallat-Lieutaud, A., & Royet, J. (2018). Cytosolic and secreted peptidoglycan-degrading enzymes in Drosophila respectively control local and systemic immune responses to microbiota. *Cell Host & Microbe, 23*(2), 215–228.

Chen, Y., Liu, F., Yang, B., Lu, A., Wang, S., Wang, J., et al. (2012). Specific amino acids affecting Drosophila melanogaster prophenoloxidase activity in vitro. *Developmental & Comparative Immunology, 38*(1), 88–97.

Chen, L., Li, Q., Nasif, K. F. A., Xie, Y., Deng, B., Niu, S., et al. (2024a). AI-driven deep learning techniques in protein structure prediction. *International Journal of Molecular Sciences, 25*(15), 8426.

Chen, Y., Zhao, Z., Liu, J., Fan, C., & Zhang, Z. (2024b). Identification, diversity, and evolution analysis of thioester-containing protein family in Pacific oyster (Crassostrea gigas) and immune response to biotic and abiotic stresses. *Fish & Shellfish Immunology, 145*, 109330.

Christophides, G. K., Vlachou, D., & Kafatos, F. C. (2004). Comparative and functional genomics of the innate immune system in the malaria vector Anopheles gambiae. *Immunological Reviews, 198*(1), 127–148.

Coates, C. J., & Decker, H. (2017). Immunological properties of oxygen-transport proteins: Hemoglobin, hemocyanin and hemerythrin. *Cellular and Molecular Life Sciences, 74*(2), 293–317.

Coscia, M. R., Giacomelli, S., & Oreste, U. (2011). Toll-like receptors: An overview from invertebrates to vertebrates. *Invertebrate Survival Journal, 8*(2), 210–226.

Dang, X., & Wang, G. (2020). Spotlight on the selected new antimicrobial innate immune peptides discovered during 2015-2019. *Current Topics in Medicinal Chemistry, 20*(32), 2984–2998.

Dudzic, J. P., Kondo, S., Ueda, R., Bergman, C. M., & Lemaitre, B. (2015). Drosophila innate immunity: Regional and functional specialization of prophenoloxidases. *BMC Biology, 13*(1), 81.

Edman, J. D., & Spielman, A. (2020). Blood-feeding by vectors: Physiology, ecology, behavior, and vertebrate defense. In *The Arboviruses* (pp. 153–190). CRC Press.

Fekonja, O., Avbelj, M., & Jerala, R. (2012). Suppression of TLR signaling by targeting TIR domain-containing proteins. *Current Protein and Peptide Science, 13*(8), 776–788.

Gueguen, Y., Herpin, A., Aumelas, A., Garnier, J., Fievet, J., Escoubas, J. M., et al. (2006). Characterization of a defensin from the oyster Crassostrea gigas: Recombinant production, folding, solution structure, antimicrobial activities, and gene expression. *Journal of Biological Chemistry, 281*(1), 313–323.

Gulley, M. M., Zhang, X., & Michel, K. (2013). The roles of serpins in mosquito immunology and physiology. *Journal of Insect Physiology, 59*(2), 138–147.

Guo, L., Tang, M., Luo, S., & Zhou, X. (2023). Screening and functional analyses of novel Cecropins from insect transcriptome. *Insects, 14*(10), 794.

Gupta, S., Bhatia, G., Sharma, A., & Saxena, S. (2018). Host defense peptides: An insight into the antimicrobial world. *Journal of Oral and Maxillofacial Pathology, 22*(2), 239–244.

Guryanova, S. V. (2022). Regulation of immune homeostasis via muramyl peptides-low molecular weight bioregulators of bacterial origin. *Microorganisms, 10*(8), 1526.

Hadjicharalambous, C., Sheynis, T., Jelinek, R., Shanahan, M. T., Ouellette, A. J., & Gizeli, E. (2008). Mechanisms of α-defensin bactericidal action: Comparative membrane disruption by cryptdin-4 and its disulfide-null analogue. *Biochemistry, 47*(47), 12626–12634.

Hao, Y., Pan, Y., Chen, W., Rashid, M. A. R., Li, M., Che, N., et al. (2023). Contribution of duplicated nucleotidebinding leucine-rich repeat (NLR) genes to wheat disease resistance. *Plants, 12*(15), 2794.

Hu, S. Y., Huang, J. H., Huang, W. T., Yeh, Y. H., Chen, M. H. C., Gong, H. Y., et al. (2006). Structure and function of antimicrobial peptide penaeidin-5 from the black tiger shrimp Penaeus monodon. *Aquaculture, 260*(1–4), 61–68.

Hu, Z., & Chai, J. (2023). Assembly and architecture of NLR resistosomes and inflammasomes. *Annual Review of Biophysics, 52*(1), 207–228.

Johansson, M. W., Holmblad, T., Thörnqvist, P. O., Cammarata, M., Parrinello, N., & Söderhäll, K. (1999). A cell-surface superoxide dismutase is a binding protein for peroxinectin, a cell-adhesive peroxidase in crayfish. *Journal of Cell Science, 112*(6), 917–925.

Jumper, J., Evans, R., Pritzel, A., Green, T., Figurnov, M., Ronneberger, O., et al. (2021). Highly accurate protein structure prediction with AlphaFold. *Nature, 596*(7873), 583–589.

Kanneganti, T. D. (2020). Intracellular innate immune receptors: Life inside the cell. *Immunological Reviews, 297*(1), 5.

Kanost, M. R. (1999). Serine proteinase inhibitors in arthropod immunity. *Developmental & Comparative Immunology, 23*(4–5), 291–301.

Kesari, P., Patil, D. N., Kumar, P., Tomar, S., Sharma, A. K., & Kumar, P. (2015). Structural and functional evolution of chitinase-like proteins from plants. *Proteomics, 15*(10), 1693–1705.

Kidwell, M. A., Chan, J. M., & Doudna, J. A. (2014). Evolutionarily conserved roles of the dicer helicase domain in regulating RNA interference processing. *Journal of Biological Chemistry, 289*(41), 28352–28362.

La Paglia, L., Vazzana, M., Mauro, M., Urso, A., Arizza, V., & Vizzini, A. (2023). Bioactive molecules from the innate immunity of Ascidians and innovative methods of drug discovery: A computational approach based on artificial intelligence. *Marine Drugs, 22*(1), 6.

Leal, E., Múnera, M., & Suescún-Bolívar, L. P. (2022). In silico characterization of Cnidarian's antimicrobial peptides. *Frontiers in Marine Science, 9*, 1065717.

Lehrer, R. I., & Lu, W. (2012). α-Defensins in human innate immunity. *Immunological Reviews, 245*(1), 84–112.

Levy, F., Bulet, P., & Ehret-Sabatier, L. (2004). Proteomic analysis of the systemic immune response of Drosophila. *Molecular & Cellular Proteomics, 3*(2), 156–166.

Li, D., & Wu, M. (2021). Pattern recognition receptors in health and diseases. *Signal Transduction and Targeted Therapy, 6*(1), 291.

Li, F. Y., Wang, S. F., Bernardes, E. S., & Liu, F. T. (2020). Galectins in host defense against microbial infections. In *Lectin in host defense against microbial infections* (pp. 141–167).

Li, P., & Wu, G. (2022). Important roles of amino acids in immune responses. *British Journal of Nutrition, 127*(3), 398–402.

Li, Y., Slavik, K. M., Toyoda, H. C., Morehouse, B. R., de Oliveira Mann, C. C., Elek, A., et al. (2023). cGLRs are a diverse family of pattern recognition receptors in innate immunity. *Cell, 186*(15), 3261–3276.

Liu, H., Xu, J., Wang, L., Guo, P., Tang, Z., Sun, X., et al. (2023). Serpin-1a and serpin-6 regulate the Toll pathway immune homeostasis by synergistically inhibiting the Spätzle-processing enzyme CLIP2 in silkworm, Bombyx mori. *PLoS Pathogens, 19*(10), e1011740.

Loker, E. S., Adema, C. M., Zhang, S. M., & Kepler, T. B. (2004). Invertebrate immune systems–not homogeneous, not simple, not well understood. *Immunological Reviews, 198*(1), 10–24.

London, N., Raveh, B., & Schueler-Furman, O. (2013). Peptide docking and structure-based characterization of peptide binding: From knowledge to know-how. *Current Opinion in Structural Biology, 23*(6), 894–902.

Luo, C., Belghazi, M., Schmitz, A., Lemauf, S., Desneux, N., Simon, J. C., et al. (2021). Hosting certain facultative symbionts modulates the phenoloxidase activity and immune response of the pea aphid Acyrthosiphon pisum. *Insect Science, 28*(6), 1780–1799.

Luo, Y., & Cai, J. (2024). Deep learning for the prediction of protein sequence, structure, function, and interaction: Applications, challenges, and future directions. *Current Proteomics, 21*(6), 561–579.

Lysak, D. H., Downey, K., Cahill, L. S., Bermel, W., & Simpson, A. J. (2023). In vivo NMR spectroscopy. *Nature Reviews Methods Primers, 3*(1), 91.

Manniello, M. D., Moretta, A., Salvia, R., Scieuzo, C., Lucchetti, D., Vogel, H., et al. (2021). Insect antimicrobial peptides: Potential weapons to counteract the antibiotic resistance. *Cellular and Molecular Life Sciences, 78*(9), 4259–4282.

Marden, J. N., McClure, E. A., Beka, L., & Graf, J. (2016). Host matters: Medicinal leech digestive-tract symbionts and their pathogenic potential. *Frontiers in Microbiology, 7*, 1569.

McAnulty, S. J., & Nyholm, S. V. (2017). The role of hemocytes in the Hawaiian bobtail squid, Euprymna scolopes: A model organism for studying beneficial host–microbe interactions. *Frontiers in Microbiology, 7*, 2013.

Miao, Z., Cao, X., & Jiang, H. (2020). Digestion-related proteins in the tobacco hornworm, Manduca sexta. *Insect Biochemistry and Molecular Biology, 126*, 103457.

Moreira, R., Romero, A., Rey-Campos, M., Pereiro, P., Rosani, U., Novoa, B., & Figueras, A. (2020). Stimulation of Mytilus galloprovincialis hemocytes with different immune challenges induces differential transcriptomic, miRNomic, and functional responses. *Frontiers in Immunology, 11*, 606102.

Nam, S. E., Bae, D. Y., Ki, J. S., Ahn, C. Y., & Rhee, J. S. (2023). The importance of multi-omics approaches for the health assessment of freshwater ecosystems. *Molecular & Cellular Toxicology, 19*(1), 3–11.

Narayanan, K. B., & Park, H. H. (2015). Toll/interleukin-1 receptor (TIR) domain-mediated cellular signaling pathways. *Apoptosis, 20*(2), 196–209.

Nosanchuk, J. D., & Casadevall, A. (2003). The contribution of melanin to microbial pathogenesis. *Cellular Microbiology, 5*(4), 203–223.

Odei-Addo, F. (2009). *Purification and characterization of serine proteinase inhibitors from two South African indigenous plants, Acacia karoo and Acacia schweinfurthii* (Doctoral dissertation, Nelson Mandela Metropolitan University).

Pal, S., & Wu, L. (2009). Pattern recognition receptors in the fly: Lessons we can learn from the Drosophila melanogaster immune system. *Fly, 3*(2), 121–129.

Park, D., Jung, J. W., Choi, B. S., Jayakodi, M., Lee, J., Lim, J., et al. (2015). Uncovering the novel characteristics of Asian honey bee, Apis cerana, by whole genome sequencing. *BMC Genomics, 16*(1), 1.

Patnaik, B. B., Baliarsingh, S., Sarkar, A., Hameed, A. S., Lee, Y. S., Jo, Y. H., et al. (2024). The role of pattern recognition receptors in crustacean innate immunity. *Reviews in Aquaculture, 16*(1), 190–233.

Piao, S., Kim, S., Kim, J. H., Park, J. W., Lee, B. L., & Ha, N. C. (2007). Crystal structure of the serine protease domain of prophenoloxidase activating factor-I. *Journal of Biological Chemistry, 282*(14), 10783–10791.

Rock, F. L., Hardiman, G., Timans, J. C., Kastelein, R. A., & Bazan, J. F. (1998). A family of human receptors structurally related to Drosophila Toll. *Proceedings of the National Academy of Sciences, 95*(2), 588–593.

Rodrigues, T., Guardiola, F. A., Almeida, D., & Antunes, A. (2025). Aquatic invertebrate antimicrobial peptides in the fight against aquaculture pathogens. *Microorganisms, 13*(1), 156.

Rowley, A. F., Coates, C. J., & Whitten, M. M. (Eds.). (2022). *Invertebrate pathology.* Oxford University Press.

Rund, S. S., Yoo, B., Alam, C., Green, T., Stephens, M. T., Zeng, E., et al. (2016). Genome-wide profiling of 24 hr diel rhythmicity in the water flea, Daphnia pulex: Network analysis reveals rhythmic gene expression and enhances functional gene annotation. *BMC Genomics, 17*(1), 653.

Schneider, J. (2022). *Characterization of the antiviral STING pathway in 'Drosophila melanogaster': Signalling and NF-κB factor activation* (Doctoral dissertation, Université de Strasbourg).

Schnitzler, C. E. (2010). *Temperature stress, gene expression, and innate immunity at the onset of cnidarian-dinoflagellate symbiosis.* Oregon State University.

Schubert, M., Bleuler-Martinez, S., Butschi, A., Wälti, M. A., Egloff, P., Stutz, K., et al. (2012). Plasticity of the β-trefoil protein fold in the recognition and control of invertebrate predators and parasites by a fungal defence system. *PLoS Pathogens, 8*(5), e1002706.

Shaik, M. M., Cendron, L., Percudani, R., & Zanotti, G. (2011). The structure of Helicobacter pylori HP0310 reveals an atypical peptidoglycan deacetylase. *PLoS One, 6*(4), e19207.

Subramanian, N., Torabi-Parizi, P., Gottschalk, R. A., Germain, R. N., & Dutta, B. (2015). Network representations of immune system complexity. *Wiley Interdisciplinary Reviews: Systems Biology and Medicine, 7*(1), 13–38.

Sun, J., Wang, L., Yang, W., Li, Y., Jin, Y., Wang, L., & Song, L. (2021). A novel C-type lectin activates the complement cascade in the primitive oyster Crassostrea gigas. *Journal of Biological Chemistry, 297*(6), 101352.

Tan, L., Liu, Y., Sun, Y., Liu, S., Lin, Z., & Xue, Q. (2025). Copper only SOD repeat proteins likely act as an extracellular superoxide dismutase in oyster antioxidant defense. *Scientific Reports, 15*(1), 20465.

Tassanakajon, A., Somboonwiwat, K., Supungul, P., & Tang, S. (2013). Discovery of immune molecules and their crucial functions in shrimp immunity. *Fish & Shellfish Immunology, 34*(4), 954–967.

Thomas, A. M., & Antony, S. P. (2024). Marine antimicrobial peptides: An emerging nightmare to the life-threatening pathogens. *Probiotics and Antimicrobial Proteins, 16*(2), 552–578.

Tiffin, N., Kelso, J. F., Powell, A. R., Pan, H., Bajic, V. B., & Hide, W. A. (2005). Integration of text-and data-mining using ontologies successfully selects disease gene candidates. *Nucleic Acids Research, 33*(5), 1544–1552.

Uengwetwanit, T., Uawisetwathana, U., Angthong, P., Phanthura, M., Phromson, M., Tala, S., et al. (2025). Investigating a novel β-glucan source to enhance disease resistance in Pacific white shrimp (Penaeus vannamei). *Scientific Reports, 15*(1), 15377.

Ve, T., Williams, S. J., & Kobe, B. (2015). Structure and function of Toll/interleukin-1 receptor/ resistance protein (TIR) domains. *Apoptosis, 20*(2), 250–261.

Vella, D., Zoppis, I., Mauri, G., Mauri, P., & Di Silvestre, D. (2017). From protein-protein interactions to protein co-expression networks: A new perspective to evaluate large-scale proteomic data. *EURASIP Journal on Bioinformatics and Systems Biology, 2017*(1), 6.

Wang, K., Zhang, H., Zhang, J., Jia, E., & Zhu, G. (2019a). Prediction of immune factors and signaling pathways in lung injury induced by LPS based on network analysis. *Saudi Journal of Biological Sciences, 26*(8), 2068–2073.

Wang, L., Zhang, H., Wang, M., Zhou, Z., Wang, W., Liu, R., et al. (2019b). The transcriptomic expression of pattern recognition receptors: Insight into molecular recognition of various invading pathogens in Oyster Crassostrea gigas. *Developmental & Comparative Immunology, 91*, 1–7.

Wicherska-Pawłowska, K., Wróbel, T., & Rybka, J. (2021). Toll-like receptors (TLRs), NOD-like receptors (NLRs), and RIG-Ilike receptors (RLRs) in innate immunity. TLRs, NLRs, and RLRs ligands as immunotherapeutic agents for hematopoietic diseases. *International Journal of Molecular Sciences, 22*(24), 13397.

Wu, Q., Patočka, J., & Kuča, K. (2018). Insect antimicrobial peptides, a mini review. *Toxins, 10*(11), 461.

Wu, X. J., Dinguirard, N., Sabat, G., Lui, H. D., Gonzalez, L., Gehring, M., et al. (2017). Proteomic analysis of Biomphalaria glabrata plasma proteins with binding affinity to those expressed by early developing larval Schistosoma mansoni. *PLoS Pathogens, 13*(5), e1006081.

Ye, S., & Goldsmith, E. J. (2001). Serpins and other covalent protease inhibitors. *Current Opinion in Structural Biology, 11*(6), 740–745.

Yokoi, K., Koyama, H., Minakuchi, C., Tanaka, T., & Miura, K. (2012). Antimicrobial peptide gene induction, involvement of Toll and IMD pathways and defense against bacteria in the red flour beetle, Tribolium castaneum. *Results in Immunology, 2*, 72–82.

Yuan, F., Pan, X., Chen, L., Zhang, Y. H., Huang, T., & Cai, Y. D. (2019). Analysis of protein–protein functional associations by using gene ontology and KEGG pathway. *BioMed Research International, 2019*(1), 4963289.

Zdybicka-Barabas, A., Stączek, S., Kunat-Budzyńska, M., & Cytryńska, M. (2025). Innate immunity in insects: The lights and shadows of Phenoloxidase system activation. *International Journal of Molecular Sciences, 26*(3), 1320.

Zhang, G., & Ghosh, S. (2001). Toll-like receptor–mediated NF-κB activation: A phylogenetically conserved paradigm in innate immunity. *The Journal of Clinical Investigation, 107*(1), 13–19.

Zhou, L., Meng, G., Zhu, L., Ma, L., & Chen, K. (2024). Insect antimicrobial peptides as guardians of immunity and beyond: A review. *International Journal of Molecular Sciences, 25*(7), 3835.

GPSR Compliance
The European Union's (EU) General Product Safety Regulation (GPSR) is a set
of rules that requires consumer products to be safe and our obligations to
ensure this.

If you have any concerns about our products, you can contact us on

ProductSafety@springernature.com

In case Publisher is established outside the EU, the EU authorized
representative is:

Springer Nature Customer Service Center GmbH
Europaplatz 3
69115 Heidelberg, Germany